THE SECRET SURVEYOR

First published 2024 by Hawksmoor Publishing

Woodside, Churnet View Road, Oakamoor, ST10 3AE

www.hawksmoorpublishing.com

ISBN: 978-1-914066-33-7

A CIP catalogue record for this book is available from the British Library.

Table of Contents

INTRODUCTION

I'm a surveyor in the UK's construction industry and have been for more than a quarter of a century. I started out even further back than that as a humble labourer on a building site, and stuck at it for about five years before I saw the light (or said 'sod that') and went to college and university to get the technical qualifications that I needed for my chosen career.

Now, I'm a senior surveyor for a national construction company, and I'd like to think that I know enough to share my thoughts – and personal insights – into this very dynamic industry.

I've seen a lot. From the sublime to the ridiculous. In fact, that's probably the range of characteristics that I experience every single working day. Some of it is great, rewarding, and enriching. Then, at other times, you feel like saying, 'Will everyone please just *grow up!*'

When you consider that, according to a 2022 report by *New Civil Engineer*, 30% of all construction work is actually re-work, i.e., putting right what we did wrong in the first place, then you can see what I mean by ridiculous. Especially when you consider that the cost of carrying out all these repairs is seven times more than the entire industry makes in profit each year.

And it affects all of us, this building malarkey, from the houses that we live in, the offices and factories we work in, the shops that we choose to visit, and the recreation spaces we might care to enjoy.

It's all around us in the infrastructure, the roads, the railways, the schools, homes, and hospitals. The cinemas, theatres, and sports stadia. The shops, supermarkets, drinking establishments, and restaurants. It is the built environment in which all of us live. It has to be constructed then maintained, serviced, and eventually demolished or refurbished. It is often re-imagined, possibly extended, and made fit for purpose for an ever-growing population and the needs of the modern-day.

That's the UK construction industry. And here are my thoughts about it. From the inside.

As some of what I say may be deemed controversial, and as I wish to comment on real-life construction projects and the companies behind them, and as I enjoy a bit of work-related gossip as much as the next person, I'll be your anonymous guide for the journey ahead.

It will be a journey right to the heart of the UK's building industry and the built environment. It will cover every facet and leave no stone unturned as we explore the fascinating world of construction.

Before I begin, though, I'd just like to add a note with regard to the prominence of male-dominated pronouns in the following text. I'm writing all of this in the first person. I'm a man. And this is my story. But all of this relates equally to females, or others – whoever you are or however you wish to be addressed.

The construction industry needs to be more inclusive and needs more people as a whole to fulfil its responsibilities to the economy and become the vibrant sector that the country deserves. So, no offence is intended to anyone with the choice of pronouns.

In this book, we'll be talking all about construction success and construction failure, the industry's undoubted achievements and also its inadequacies. In other words, why it sometimes goes right and also why it often goes wrong. This is the inside story on the big and the small, from your new kitchen, bathroom, or home refurbishment, all the way up to the new shopping centre or the new motorway which is about to blight or enhance the environs of your home town. Basically, how it all affects *you*.

For an industry stacked with experts, with much of its activity governed by regulations, we'll be asking why we build new homes on floodplains, charge ground rent on homes 'sold' to unsuspecting first-time buyers, wrap high-rise buildings in flammable blankets, and award billions of pounds-worth of contracts to clearly-failing companies. We've even seen the whole of the top brass at the RICS resign after a recent major scandal.

So, don't leave home without reading this guide.

In fact, don't even *stay* in the house without reading it first!

I am The Secret Surveyor

1. THE UK CONSTRUCTION INDUSTRY

AN OVERVIEW OF THE INDUSTRY

Construction is the third largest sector of the UK economy. It employs almost three million people, approximately one-tenth of the UK's entire workforce. It contributes close to £200bn annually to the country's GDP. It affects, literally, everybody living in the United Kingdom.

Britain is the most populated country in Europe. We, the people, live cheek by jowl. Those green fields and idle pastures that you might pass in the car, or on the train, are but an oasis.

Cities, infrastructure, and the services that link them together are apt to propagate. In the not-too-distant future, all of that greenery will lessen and reduce, and all of those concrete apparitions will grow and eventually merge.

People don't change. From time immemorial, humans haven't really evolved all that much. They need a water source; hence, most population centres are built near rivers. There, the inhabitants can wash themselves, fill pots for the purposes of cooking, scrub their clothes, and flush away their ablutions.

A wood or a forest nearby is good. It provides fuel for the fire, plus the building materials for walls and a roof to your shelter or accommodation. These dense, wooded areas have also traditionally been a good place to run to and hide in should invaders come marauding. People, throughout history, don't really change.

But technology does.

Who invented concrete? Someone did (around 3,500 years ago).

Steel? Someone again (a precursor to modern steel dates back to c.1800 BC).

We can build higher, better, and longer-lasting. We can build curves. The Guggenheim, anyone?

From the wonder that is the ancient pyramids up to the magnificence and opulence of the Burj Khalifa hotel. From Bronze-age mud-brick constructs, via expertly crafted stone masonry, all the way up to modern-day auto-cleaning glass facades, all we ever really do is house people and serve the needs and wants of the human beings living in a particular era.

The UK has a massive housing shortage at present. We simply can't build accommodation fast enough. With an ever-increasing population, it is difficult to see how the gap between what we have and what we need will ever be successfully bridged. So, for housebuilders – the largest of whom are the Barratts, Taylor Wimpeys, and Redrows of this world – it seems as if they have a license to print money.

These developers or main contractors, while employing their own staff – consisting of admin and office people, management and professionals, craftsmen and apprentices – still rely on a whole array of subcontractors. They will give work to groundworkers, bricklayers, joiners, roofers, plumbers, plasterers, electricians, painters and decorators, and specialist trades.

They will place orders for materials with suppliers and builders merchants. The money spreads across the whole of the economy. Those builders rock up to a café across the road from the site and enjoy a hearty breakfast and pay for it. They spend their wages on the weekly supermarket shop, or down the pub, or on clothes and school uniforms for their kids.

The revenue derived from construction activity seeps into every pore of the UK economy. It is earned, and then it is quickly spent.

Even better, the product that it builds – the home that it sells to you – is likely to go up in value the longer you own it. It's your house and you love it, but it is also a valuable asset. And an appreciating asset at that. And how often can you say that about whatever it is that you choose to spend your money on?

The construction industry flows through the very fabric of the entire country. It builds, renovates, restores, and regenerates the towns, villages, and cities in which we all live. It then links them all together via an infrastructure of roads, railways, harbours, and airports.

It feeds them with gas, electricity, water, and the cabling for technology. It builds bridges and tunnels, dams and reservoirs, tourist attractions, leisure facilities and sports centres. Cinemas, libraries, as well as hospitals. Schools, offices, and warehouses. From a detached garage at the side of a house up to a statement-piece skyscraper, the construction industry gets it done.

It pays its taxes and it shapes our world – one brick, column, or single-glazed panel at a time.

And the good thing is this, the majority of this country's construction activity does not cost the British taxpayer a single penny. *How so?* I hear you ask. Well, please allow me to explain.

Approximately two-thirds of construction output in the United Kingdom takes place as a result of private housebuilding and private commercial enterprise. That is, businesses, entrepreneurs, developers, and financiers taking a punt on projects in which they hope to see a tangible return on their investments.

Social housing and government-funded schemes, while often grabbing the newspaper headlines, do not make much of an impression on the overall industry outlay. Infrastructure makes an impact, especially with every road in the country seemingly buried under an avalanche of traffic cones at the moment. Still, even expenditure on infrastructure is dwarfed by private sector work at the present time.

No one can predict the future. No one can say what the country will look like in 10, 20, 50, or 100 years' time, but whatever people need at that particular point in their lives, it will either be built from scratch or it will be adapted from our current stock of buildings.

The construction industry is huge. Yet it is only third on the list of importance for economic impact. Retail is bigger. We shop and we need to eat. In pounds, shillings, and pence, the shopping market generates more money per head of population than any other industry in the land. This is followed – apparently – by the banking and pension market, which is the second biggest money-making industry in the UK in terms of turnover.

These are obviously both important industries. They are bigger in terms of monetary revenue, but with regard to the actual physical contribution to our world, construction wins it for me. It builds Britain's landscape.

And what is it that the industry aspires to do? Here are the mission statements of some of the biggest construction players currently on the circuit.

- Arcadis: Improving quality of life.
- Vinci: Forging a sustainable future.
- Wates: Creating tomorrow together.
- Morgan Sindal: Creating a better built environment.
- Sir Robert McAlpine: Proudly building Britain's future heritage.

Forging and creating. It's like art, a landscape, where someone paints a picture that we can all appreciate. Someone constructs a building that we can live in, visit, or admire as we walk on by and go about our daily business.

It is manifest. It's there. It occupies a space.

In turn, this is an industry much-regulated by government. There are planning laws, building regulations, initiatives, and reports, all of which are meant to improve the ability of the construction industry to perform and deliver.

Many of these initiatives are simply wishful thinking, in my opinion. I don't know if the authors of some of these reports have ever even set foot on a building site.

This is a volatile industry. If you are lucky (and sometimes unlucky) enough to work within its walls, you'll know on an almost daily basis just how volatile it can be.

Try telling the manager of a bricklaying firm, who's had 15 of his lads working on your building project for the past month, that you haven't got the money to pay him. You'll hear volatile. You may even feel it!

And it's a volatile industry in other ways, too. When money gets tight, what do you spend yours on?

I bet the planned holiday and the new set of clothes you want are the first items to be ditched.

When push comes to shove, you'll spend your last pennies on food. Last luxury? Loo roll.

You won't build that home extension. Your local council won't build that new school or that new public library. The construction industry will suffer. Its employees will disappear and find whatever jobs they can in order to feed their families and pay their bills. Some of them, even when the market returns, will be once bitten and twice shy. If they've found a steady earner in the meantime – cab driving maybe – they might stick with that. Their practical skills will be lost to the building industry forever.

But I still maintain that it's a great environment in which to work when things are going well.

In 2020, while the UK entered a lockdown caused by the Covid-19 outbreak, most of the country's building work continued. Somehow, construction activity appeared to be exempt. Maybe no one had the balls to tell a muscle-bound workforce that they had to stay at home.

Construction work did not stop and, on the contrary, the pandemic period became one of the building industry's most productive spells. Everyone was busy; even cowboy builders. The good ones had (and continue to have) a full order book for months, if not years, to come. Try to hire a decent building contractor right now and you'll join the back of a very long queue.

Material prices also went through the roof. Across the board, in the past couple of years, a realistic estimate would be a minimum 25% increase (annually) in the cost of steel, bricks, plaster, plastic, timber, glass, etc.

At the same time, the most recent government initiative asked for construction costs generally to be reduced by 33% for almost each and every activity. In other words, we should aim to build things for quite a lot less. If material costs have risen by at least 25% in a 12-month period, the only way to produce such a saving is to effectively halve the wages of everyone in the industry. Wages, by the way, have also seen a similar steep rise.

That government report, called *Construction 2025: Industrial Strategy* also asks for a 50% reduction in the time taken to complete projects. I'm actually speechless on what to say about that. Work faster and earn less. Only a government could come up with such a ridiculous idea.

Read further into that same government initiative narrative and you'll find that they think these savings can be achieved by better programming and the use of 'offsite production', as if the industry is suddenly going to have a lightbulb moment and realise what they've been doing wrong for all these years.

I'm all for innovation. I'm all for labour saving devices that free people up to lead better lives, but it makes no sense to just say we should be able to do things for half the price using the same materials and the same pool of workers that we currently have. How? I mean how? You can't alter the reality of a large, important industry with just the aid of a few soundbites.

Construction changes the landscape and the environment. Retail can't do that. The pension and banking industry can't do that. Though third on the list for so-called importance to the UK economy, I think we may make the most valuable contribution of all.

The stuff that we do is important.

And it's likely it will always be thus.

THE HISTORY OF THE CONSTRUCTION INDUSTRY

For hundreds and even thousands of years, nothing much changed in the living habits of our ancestors. Then, with the coming of the Industrial Revolution in the late 18th century, it became possible to mass-produce items that could be used within the very fabric of our homes and put to good use in our everyday lives. Think concrete, steel, glazing, and masonry. The building blocks of the society that we recognise today.

I remember when I was a student at university, doing a module in Building Technology as part of a surveying course. We were in the lab one day, and the tutor showed us some bricks (exciting, I know), then asked if anyone knew what type of bricks they were.

No one answered.

Okay, we were still learning at the time, but it's sad to think that those young men and women, unable to answer such a simple question, are now generally the senior managers in the industry!

Eventually, I offered a response. 'Fired clay?' I suggested.

'Well, yes,' he said. 'We stopped using mud bricks a long time ago!'

The answer that he was actually looking for was 'facing bricks' – the type that generally adorn the outside (i.e., the face) of buildings. It's also possible to have common bricks (less attractive and usually buried in the ground or out of sight), and engineering bricks (solid and usually used in foundations).

The point is, we once used mud bricks. Then, when we became industrialised and had developed the technology to do so, we could advance and make a superior quality product, and in far greater quantities than ever before.

We evolved. The industry evolved.

British engineer John Smeaton invented what might be termed *modern* cement in 1756, which subsequently led to the development of Portland cement (a product that could be poured while still wet up to great heights, and which would then cure and set like, well, concrete).

In turn, in the 19th century, the Bessemer Process, named after Sir Henry Bessemer working out of Sheffield, who patented his idea, was the first cost-effective system for the industrial mass-production of steel.

In other chapters from the Industrial Revolution, machines were developed that could turn out vast numbers of bricks, tiles, building blocks, and any number of items that were useful for construction.

Those little clay tiles that you see on some roofs (usually the ones on nice houses, in suburbs, the countryside, or in fairly posh neighbourhoods), well, they used to be individually handcrafted. Then somebody developed a mould and poured wet clay into it, still churning each tile out by himself. Nowadays, they are all made by machine, although some still bear the hallmark 'handcrafted'. Do you know what this means now? As they roll off the assembly line, someone in production has the task of simply touching them. The merest touch of a finger or thumb enables them to then be labelled 'handcrafted'. They're not – not like in the old days – but, then, I suppose we don't really want those days back anyway.

Nowadays, we can build higher, longer, stronger, weirder, and wider than ever before.

The Shard is the tallest building in Britain. Its tapering shape is unusual and distinctive, and the glass façade is constructed from more than 11,000 glass panels that cover an area of over 56,000 square metres. The total floor span is over 31 acres, and all located in the heart of Central London. And the best thing about it? 95% of the construction materials used in the build were recycled.

Today, we live the lives we largely want or the ones that we have created. And the construction industry, evolving from the old ways to the new, has played a big part in that human journey.

If you've ever read Ken Follett's *The Pillars of the Earth*, you'll get a good idea of how things used to work. This particular story is set in the 12th century, and the client is the local archbishop. He wants to construct a cathedral, using funds obtained from his parishioners (given none too happily in the form of taxes, one can imagine), and so appoints a master-builder to manage the project.

In such times, the master-builder would act as the architect, drawing up the plans, and also as project manager, piecing the different elements of the construction together. He oversaw the build phase right up to completion. He might even get his hands dirty laying many of the stones himself and crafting the more intricate parts of the building. He would have a host of workers quarrying and delivering the stones, and skilled craftsmen and labourers then building the envisaged construct from the ground up.

My point is that in charge of it all was *one* man, with many responsibilities.

By the time we get to the industrial age, we find a more entrepreneurial kind of operation in action.

I love the story of Joseph Williamson, who walked from Warrington to Liverpool as a 14-year-old youth and found employment with a wealthy tobacco merchant called Richard Tate. Young Williamson prospered in Mr Tate's tobacco business and then married the boss's daughter (never a bad idea), eventually inheriting the family firm when the old man retired.

Seeking a suitable marital home for himself and his new bride, Joseph bought a large plot of land on a hill at the edge of the city, appropriately called 'Edge Hill'. He found sandstone in abundance in the ground all around him. For labour, he decided to give employment to the many ex-soldiers who were to be found living on the streets at that time. These were veterans of the then-recent Napoleonic wars. Many either injured, traumatised, or merely destitute.

Williamson put them to work building a large detached home for himself and his beloved wife, Elizabeth.

Then, once the house was complete, he found his workforce turning up on the following Monday, caps in hands, looking to see what else their benefactor might be able to offer them.

Suddenly finding himself responsible for this hungry but eager army, Joseph Williamson put them to work building a row of terraced houses on the adjacent plot to the land he had purchased to build his new home.

The houses were built and Williamson was soon a landlord with properties to rent.

After completion of the new row of houses, the men turned up at the boss's door once more. Williamson scratched his head. 'I suppose,' he might have said, 'that as we are on a hill, let's build a raised platform around the back of the houses to give these properties a garden.' From these platforms, he then built tunnels to the road down below for ease of access for his new tenants.

The tunnels eventually became a network. The skills that his employees developed, mastering the art of brickwork – building tunnels, archways, and structural supports – would eventually gift to the nation a workforce with the ability to help build Britain's emerging railway system.

And the best thing about the entire enterprise for me is that it was born out of altruism. When the tunnels themselves were being built for mere folly, the man behind it all kept that fact a secret to himself. He encouraged his workers to slow down, to take pride in what they were doing, to master the techniques involved, and get inventive, creative, and to ultimately become formidable in their talents.

He not only gave these desperate people bread for their tables, he also gave them their dignity. The Williamson Tunnels are now an award winning tourist attraction, open to the public. This is truly the gift that keeps on giving.

Of course, the industrial age gave us many such people of similar vision and fortitude although, nowadays, we no longer rely on the ability or vision of one person to carry so much of the burden. We have contractors and subcontractors, clients and architects, and a so-called holistic approach to construction. Everyone on the same page, pulling in the same direction.

If only that were always the case. One could write a book about the travails of the modern construction industry. And now I have!

WHAT THE CONSTRUCTION INDUSTRY DOES

The housing market is an obvious example of what the construction industry does, since it is responsible for more than a quarter of all of the UK's construction output. This covers everything from one-off private houses, to private developers buying land and building on it, right up to the large household-name building companies who buy up land and then build annexes and adjuncts to existing towns, and even (on occasion) whole new towns themselves.

Moving on, the next largest sector of construction activity is in infrastructure. This is the connectivity that filters through to each corner and into every nook and cranny in the land.

Think bridges and tunnels, roads, motorways, railways, airports, harbours, and large-scale power plants. And don't forget the services that supply us with our energy and other needs. I'm talking about gas, electricity, our water supply, and IT cabling. Digging up roads and pathways, installing the ducting and the pipework, it's all part of the United Kingdom's supporting infrastructure.

And then there is the education system, which is responsible for around 10% of all construction output. Schools, nurseries, colleges, and universities, popping up wherever people need them, nourishing the younger generation, and providing jobs for teachers, head teachers, and support staff, including cooks, cleaners, school caretakers, and the like.

Much of our current stock of schools no longer meet the needs of their ultimate clients, our children. Technology moves so fast, and no one has a better handle on it than the kids, that schools have to continually improve to keep up with the demands and expectations of the next generation. That means either building new schools or making extensive alterations.

This is a large market for construction. Especially during the summer recess, when schools are closed and allow building work to take place unhindered.

Another big area for the building industry is in the commercial and retail sector. Your shops, supermarkets, and malls that make up a further 12.5% of construction work.

In turn, hotels, leisure, and sport come in at around 9% of total industry expenditure. The health and medical sector represents just 5% of construction activity, although those new hospitals and health centres do not come cheap.

Industrial schemes, including warehouses and factories, make up a further 11.5% of industry turnover, although this figure is growing all the time. Online retail means mass distribution, which means more warehousing and an expanding market.

Construction also involves the services to all of these things, and the infrastructure in between.

This is what construction does. It changes the landscape.

WHO IT EMPLOYS

Who works in the construction industry? Well, human beings from all walks of life, all sorts of characters with all sorts of skills.

Indeed, there are some very funny characters at work. I love the story about the MD of a national roofing company who sent a round-robin email to his staff to announce the departure of his office manager, a very demanding woman who was not liked around the workplace. In fact, she was considered a bit of a cow. The MD said, and I quote, 'She has decided to move on to pastures new. No pun intended!'

Anyway, want something knocking down? You can get a gang of lads from the pub on a Sunday to (hopefully) turn up to site on Monday to take something apart by brute force and then chuck it into a skip. No real qualifications needed for that one. That's the demolition guys for you.

Similarly, you can have an onsite labourer – just a humble fetcher-and-carrier – who will keep the workplace tidy and do a bit of brushwork and shovelling and empty out the bins. Good honest work. And always remember, *a tidy site is a safe site!*

A bit of a step up from the labourer is the craftsman's mate. He'll still fetch and carry, but with a bit of nous thrown in. He's still doing some of the lifting and lumping, but he knows what he's doing and why. Always keeping the craftsman working, that's where the money is. That's the bit that we all see: the finished product such as the wall built or the plastering done. The time-served tradesperson doesn't want to mix their own plaster or cement, or fetch the bricks or carry the plasterboards to where they are needed. They want a hand with the heavy lifting, and that's where their mate comes in. A semi-skilled person. Essentially, it's labouring with an added bit of wisdom.

Then there are the tradesmen and tradeswomen. Your plumbers, plasterers, electricians, joiners, brickies, roofers, painters and decorators. They've taken the time to learn their craft. They can cost you a pretty penny, and the good ones are *always* busy. You can appreciate the results of their endeavours. These guys and girls produce, and their efforts are what you see in the finished article of your construction.

But there has to be a plan, right? And someone has to manage that plan to fruition. Hence the need for site managers, working foremen, architects, designers, building surveyors, quantity surveyors, project managers, contract managers, commercial directors, planners, buyers, sales agents, legal support, estimators, and the admin staff and the company directors who tie it all together.

There are hundreds of different roles within construction, and almost three million different people that do them; all of them individuals, and each one of them part of a dynamic industry.

With the dying out of the traditional apprenticeship schemes in the last half century, there is now a significant shortage of skilled men and women. The industry has a shortfall of both labour and professional personnel whilst demand is set to increase with more than a quarter of a million more construction workers needed in the next five years alone (projected up to 2028).

Here's a brief breakdown of the construction jobs market and who it currently employs.

- Groundworkers: 3%
- Bricklayers: 6%
- Steel erectors: 2%
- Plant drivers: 2%
- Cladders/envelope: 4%
- Glazers: 3%
- Joiners: 9%
- Plasterboard/drylining/interior fit out: 6%
- Floorers: 2%
- Roofers: 5%
- Plumbers: 10%

- Electricians: 10%
- Plasterers: 3%
- Painters and decorators: 6%
- Scaffolders: 2%
- Labourers: 7%
- Site managers/Supervisors: 3%
- Specialist trades: 2%
- Management and professionals (e.g., architects, surveyors): 15%

It's encouraging for me to see that roughly twice as many people are employed to manage, orchestrate, and direct a project as there are to just lump the gear from one place to another onsite.

I think it's important to have experienced personnel in charge. Someone needs to be responsible and also to be *held accountable* for ensuring that everything goes (as near as damn it) to plan. That's what they get paid the big bucks for. We should make them earn it, since with great power comes great responsibility.

But there's room for pretty much everyone in this great industry of ours. We can all play a part. If you're a young person entering further education or the job market, or if you simply fancy a career change, you could do a lot worse than considering construction. The opportunities are diverse, the work is interesting and rewarding (or at least it can be), and it could provide you with a decent income for many a good year to come.

Barring a recession, of course.

I've seen a couple of those in my long surveying adventure. The early 1990s were particularly bad. There was very little work around, and even schemes that were then active suddenly ground to a halt. Projects that were still at the planning stage were simply mothballed.

House prices crashed, effectively bringing a stop to work in that sector. According to an article in *Construction News* in December 1990, building contractors faced a £3 billion nosedive in activity and output for the following year alone.

I spent that time in college and university and by the time I emerged – suitably qualified – the industry was hiring again. Good timing for me.

In the late Noughties, it happened once more. Scores of applicants for every vacancy and I only just managed to ride out that particular storm. There has been constant work ever since.

No industry is recession-proof, and everyone has to do what they must in times of crisis in order to pay the bills. But it is far better to have a skill or a trade that you can turn your hand to, rather than not have one, in my opinion.

And, to be honest, a lot of what gets built in the world of construction is less than perfect. The people who build these things often focus more on speed than quality. So, the end result won't last forever. These buildings will need to be maintained, and many will get knocked down and rebuilt in 20-, 30-, or 40-years' time. If you stick around in the industry long enough, you might even get paid in the future to go back and correct your own work!

Construction is an industry that just keeps on giving.

ECONOMIC SIGNIFICANCE

How important is the construction industry to the UK economy? I would say that it is huge. It is in the top three sectors on Great Britain's balance sheet, it accounts for about seven percent of the UK GDP, and it *directly* employs 2.1 million of the 3 million people listed previously.

But this is only half the story. These figures for construction do not include the professionals that serve the industry. The contributions of architects, surveyors, engineers, and other management personnel are included within the figures for a separate sector: the service industry.

Meanwhile, production of all the materials used in the building industry is catalogued under manufacturing. The place that makes the bricks? That's manufacturing. The plasterboard? Manufacturing. You get the idea. Their names do not appear in the credits for construction.

Similarly, with the suppliers, and the builders' merchants – the Travis Perkins, Wolseleys, and Huws Grays of this world – they all come under retail.

All of those enthusiastic DIYers? They are discounted in the official employment figures. All of their hard work is as nought when it comes to the counting. The contribution of construction to the GDP figures is based purely on physical, actual site construction; what happens behind the scenes – in factories or in builders' yards – is not included in the data.

According to the Chartered Institute of Builders (CIOB) 2014 report entitled *The Real Face of Construction*, the actual contribution that the building industry makes to the country's economy is about double that of the figures served up by the Office of National Statistics.

So, we're talking about approximately 15% of the UK economy in real terms.

Like I said, it's pretty big. And not only is it important in its own right, it also provides the physical assets and the environment for many other sectors to strut their stuff.

The high street, where we still do much of our shopping, is little more than a collection of buildings. Prefer to visit that edge of the city retail park? That's nothing but a big bunch of industrial units. Like to shop online? All of that merchandise is stored in a huge warehouse, which is nothing more than a steel-framed, metal-clad building.

Oh, and how does your purchase eventually find its way from the warehouse to your home? Via the infrastructure, the roads and the motorways that were built by construction.

In the last hundred years, construction's contribution to Britain's economic output has barely changed as a percentage of the overall GDP. While agriculture has declined significantly, and manufacturing likewise, construction has been consistent. If, as the old saying goes, we are a nation of shopkeepers, then I think it's fair to say that we are also a nation 'on the tools'. We're craftspeople; we're a nation of builders.

To underline the importance of the building industry to the UK economy as a whole, I think the argument is best made when we consider that three-quarters of our nation's capital wealth – its actual valuable assets and its tangible worth – are measured by its physical properties.

I am talking about the 28 million homes in Great Britain, and the 24 million other types of building, whether they be hospitals, offices, shops, or schools. There are 46 working airports. There are more than 500 ports and harbours of various sizes and descriptions on our shores.

The coastline of mainland Britain, together with that of Northern Ireland, comes to 15,747 miles, give or take the odd coastal landslide. Divided by the number of ports and harbours (actual figure of 520), it means that there is a port – whether that be passenger, cargo, or a combination of both – every 30 miles. Perhaps, then, we are also a nation of seafarers and oceangoers.

Construction also has the power to add value to land. A field is just a field, but award it planning permission for the building of a dozen new homes and its worth increases exponentially. That land becomes a desirable asset, and increases in value again when those 12 homes are built. What was once an empty greenfield site, worth £1 million per acre, is suddenly worth quadruple the price, less the build cost of the houses. You can double your money as easy as anything.

As the American author Mark Twain once said, 'Buy land. They're not making it anymore!'

There are many ways in which construction contributes to the economy. For one, it is truly egalitarian. With so much current talk of 'levelling up' in the UK, construction is one of the most evenly-spread industries in the land.

There will always be a glut of projects in London and the South-East region of England – we can't deny that that's where the majority of the money is – but there are as many people employed in construction, certainly percentage-wise, in every other region of the country as there are in the more prosperous South.

Added to more than 2.1 million workers directly employed within the building industry, another 600,000 work within the services that support construction. Three hundred thousand more are tasked with manufacturing the products that those builders need on a daily basis.

That's three million people. That's an economy right there. And, as we've already seen in the CIOB report, that's only the half of it, as many construction-related activities aren't included in the figures because their work does not take place on an actual construction *site*.

To break these numbers down further, those two million-plus contracting jobs are employed within a total of over 200,000 construction companies. The average company size then, for the whole of the UK, is ten or so employees per business. This must mean a lot of these guys and girls are self-employed, on the tools, doing the actual (physical) work of construction.

Within the number employed in the services sector, those 600,000 are to be found working within about 30,000 companies. Average employee count in that sector is, therefore, 20.

Within the manufacturing sector, those 300,000 are employed in 20,000 companies. Average employee count in that arena is 15.

Lots of people. Lots of businesses.

That activity generates income and expenditure. Not to mention the VAT. It all adds up, it all makes a difference, and it all goes into that pot of money that we call the UK economy.

2. THE DIFFERENT SECTORS

THE HOUSING MARKET

Houses. They come in all shapes and sizes, and we all tend to live in one of one kind or another. Of course, there are an unfortunate minority of people who are homeless, and the odd Jack Reacher-type out there who wanders the earth, like Caine in *Kung Fu*, righting wrongs when they find them – 'wherever I lay my hat' kinds of characters.

But let's get back to houses and the existing stock of accommodation in the United Kingdom. I'm talking about the individual dwellings that make up the towns, villages, hamlets, and major cities that stretch the entire length and breadth of the British Isles.

If you have the money to purchase this book, and the time and the space and the security in which to sit down and read it, then you probably live in a house (or an apartment). It might be a terraced home or a detached, semi, or bungalow.

It probably has four external walls and a roof. Windows too. A garden, yard, or some sort of outside space if you're lucky.

Your house fulfils a primary function for you. First and foremost, it satisfies the need for shelter. Protection. Security. A base. A place to call home.

If you're lucky enough, you might even own that home. Or, you might think that you own it, but you're actually still paying your mortgage, in which case, let's see who *really* owns your home should you default on the payments. (I sincerely hope that never happens to you.)

You might rent from a private landlord or a social housing provider.

However you roll, in that house of yours, we pretty much all need, want, and have a place to live.

In the UK, we also have newbuild housing schemes taking place right across the country. At the edges of our towns and cities, farmers and other landowners are selling off available plots by the acre or hectare, and – in the eyes of many – the British countryside is disappearing under a tide of bricks and mortar.

Where land is in short supply, such as in central city areas, old warehouses, factories, and redundant buildings are being snapped up by developers and converted into living spaces.

As many readers will undoubtedly be aware, the UK currently has a massive housing shortage. Due to immigration, an ever-increasing population, plus social changes (e.g., more single-person living), the demand for more housing stock is intense. In a nutshell, we simply *need* more places for people to live.

The government is trying to meet some of that demand. The recent *Help to Buy* scheme was one such example, running for a decade from 2013, and taking final applications in late 2022. Over 300,000 people bought a home with the help of the scheme, with more than 80% of those being first-time buyers. If you had 5% of the mortgage value of your desired property, and could persuade a bank or building society to mortgage a further 70%, well, the government would make up the difference for you.

But only if you wanted to buy a newbuild house.

It was great while it lasted for housebuilders like Persimmon, Barratts, and Taylor Wimpey. They had a pool of people who could easily purchase their products. But, what if you didn't have a penchant for newbuild houses? Well, for that particular scheme, you were out of luck.

Not everyone *likes* a newbuild. Not everyone *should.*

You know the old saying, 'They don't build them like they used to.' Well, in housing, I would say that this is essentially true.

Look at a row of terraced housing built a hundred years ago. Notice the detailing around the windows, the corbelling at the eaves, the artistry in the brickwork. Little bits and bobs that stand out and reward the eye and the time that you take to study them.

Now take a look at a modern newbuild house. Notice the detailing? No. Why not? Because it's not there. It's flat. Uniform. Mundane in many ways.

An old university lecturer of mine used to write a weekly column about construction for a national newspaper. In one such piece, he talked about how he had just finished playing football and noticed the way the sweat running down his forehead was running off his eyebrows. The water was being dispersed, diverted from his eyes to allow him to still see where he was going. His eyebrows were an evolutionary human device designed to protect his vision.

He then noted how many old buildings employed a similar methodology.

Architects and craftsman-builders essentially put 'eyebrows' on their buildings. Rainwater that ran down the façade of a building would encounter a 'cill' or some other overhanging feature, and this would then propel the rain away from the exterior fabric of the building. That was because – if you let the external masonry get wet – the inside of the building itself would become damp. And damp has a way of penetrating right through to the interior which causes the people inside to get ill. Bronchial problems, particularly in the children and the elderly. So why not build it right? Stop the problem at source before it ever becomes an issue.

Nowadays, developers are content to build the featureless home, with no more overhangs. When the walls get wet, conventional thinking is that the central heating will probably dry it out. At huge cost to your bills, of course, and at the expense of the environment through global warming or climate change or whatever you want to call the heating up of the planet.

So… old housing stock, new housing stock – there's a difference.

But we need more houses. The UK housing shortage is currently said to be between one million and two million dwellings, depending on which report you choose to read or believe. Whatever the case, no one denies that there *is* a shortage, and that it is of significant proportions.

In 2021, 200,000 new homes were built, yet experts estimate that Britain needs 300,000 per year to be built – for the next ten years – just to catch up and address the smaller of the deficits in those two estimates. That's good news for housebuilders. If they build it, you will come. Those houses will sell.

And as for the quality? Well, I stick with what I said. They don't build them like they used to. Whilst I could fill another book with examples, I shall resist the urge and simply whet the appetite.

A December 2021 article on the This is Money website into *New Build Nightmares* reported: "About 94 per cent of newbuild homeowners report at least one defect once their property is complete, according to a recent poll by trade body the Home Builders Federation (HBF). Poor plastering, bad brickwork pointing and damaged windows are among the most commonly reported 'snags'. The average property now comes with as many as 157 defects, up by 96 per cent from 80 in 2005, according to specialists BuildScan."

Meanwhile, in October 2022, the Daily Mail reported on a couple who spent £450,000 on a newbuild in Northamptonshire only to identify more than 250 faults, including "collapsing walls, broken doors, and dodgy electrics."

And what about when you shouldn't build them? For example, when you don't have planning permission, or when the land that your proposed development will sit on is contaminated. Such is the case at the moment, where a company called Countryside Partnerships recently built 263 new homes in Cheshire without first satisfying the pre-planning conditions – namely to de-contaminate the land – before sticking a spade in the ground.

They pressed ahead anyway, eager to build and then sell their houses. Now, the new owners are unable to insure or sell or re-mortgage their new homes.

At the time of writing, a retrospective planning application has been lodged. Worst case scenario, it won't be granted and the houses will all have to come down. Best case scenario, the land will be found to be fit for purpose after all and life – currently on hold for the residents – can go on.

But these developers shouldn't take anything for granted. They have been accused of arrogance to the point of contempt. Another pre-planning condition, similarly ignored, was that they preserve part of a historic wall on the site. That didn't happen either and, sadly, it appeared to just fall down. When one local councillor heard the excuse that the supposedly protected wall had fallen down, he asked if Humpty Dumpty had sat on it!

Efforts are also being made to build in particularly tricky areas elsewhere. London Mayor Sadiq Khan, for example, was given almost £5 billion of government money in 2016 to build 116,000 affordable homes in the Capital before 2023. By the first quarter of 2023, he had achieved almost 80% of his target, and due to a reboot of the plan, he now has until 2026 to spend a further £4 billion.

There's no doubting the demand for the finished product. After all, London has one of the most severe housing shortages in the country. But my point is that even the mayor, with all of that money at his disposal, still can't find the people to build them fast enough.

I'm sure, given how crowded our capital city is, he probably can't find the space to erect all of these new properties anyway, even if he could find the construction companies to build them.

So, we build hundreds of thousands of houses each year, and it's still not enough. Equally importantly, how do we deliver the services required for all these new properties?

If you live in a city, can those old underground sewers – that still do much of the dirty work – take the strain of all these additional houses that are being built on every patch of available land?

In turn, when you buy a new house, there will be sinks, toilets and showers, and that all means running water. Whilst it rains a lot in the UK, we capture barely a drop of it, percentage-wise. Water shortages are becoming more common, so if we want to build more houses, we are going to need more reservoirs. We need to capture that precious resource when it falls from the sky.

When it comes to building houses, it's about adequately servicing them, too.

THE HEALTH SECTOR

Since the foundation of Britain's National Health Service in the 1940s, which was a gift to the nation after their formidable efforts in the Second World War, the general public has come to expect that the NHS will be there for them in their hour of need.

Despite record hospital waiting lists, no doubt exacerbated by the Coronavirus outbreak that landed on our shores in early 2020, most of us still like to believe that we have a functioning national health service.

And it is only right that the people working in healthcare should have adequate facilities in which to perform their important and often life-saving work.

Sadly, that is not always the case.

There is a shortage of beds, and patients can be left waiting in corridors or reception areas for hours or even days at a time. The maintenance of the buildings themselves is leased to the lowest bidder, with obvious consequences in terms of overall standards and staff motivation.

Hospitals, surgeries, and health centres are specialist establishments. They need attention to detail in their design. They are highly functional, and staffed by highly skilled and rightfully demanding professionals. And yet, almost half of Britain's current medical facilities are no longer deemed to be fit for purpose.

For the time that any new healthcare units are under construction, the sector makes a valuable contribution to employment in the building industry. While the internal fit-out stage, towards the end of the project, might see specialist trades involved, the majority of what goes on beforehand is just bog standard stuff. It is brickwork, plastering, painting and decorating. It is joinery, glazing, and nothing very out of the ordinary.

The work gets finessed the closer it gets to completion, and hopefully, everyone has been following a well-thought-out plan from the start, but construction work in the health sector is just a slightly more complicated version of the usual building model.

Except it's more expensive. Because the taxpayer is paying for it.

There are numerous examples of the outrageous overcharging that appears in the public as opposed to the private sector. How about £302 for an electrical socket (about ten times more expensive than the top-of-the-range item at Screwfix), or £5,500 for a sink (probably 20 times more expensive), or £25,000 for three parasols (about 50 to 100 times more expensive).

And in the Health sector, there are a lot more stakeholders involved. There are consultants to consult with, as well as teams of doctors and surgeons asking if they can have this, that, and the other. And it all costs. A lot.

And when a new hospital is planned, a contractor is appointed, and that contractor then goes bust, the aftereffects and the spiralling costs can be felt for many years.

This was the case when contractor Carillion went bust in 2018. The local authorities in the West Midlands and Liverpool were both due to mothball old hospitals at the time, with a seamless transition into shiny new facilities for both patients and staff envisaged as part of the process.

Cue administration. Cue confusion bordering on chaos. Cue £7 billion of debt.

In the aftermath, there were only a handful of construction companies capable of rising to the task of putting these problem projects right. And guess what? They weren't going to come cheap.

There's little worse than taking over someone else's half-built construction project. First of all, you have to check the quality of everything you're taking possession of. After all, you're now assuming liability. This process can take months or even years to establish. Meanwhile, the old, crumbling facility has to be maintained beyond what was deemed to be its realistic end date. That's another massive cost right there; just keeping it up and running. Then, the new company can give a take-it-or-leave-it price to complete the building work.

Sound like good value for money? You bet it ain't. But, to be fair, it's someone else's mess. Why should any contractor sell themselves cheap to come to the rescue? I bet Red Adair charged a fortune. And he was worth every penny.

And when a subcontractor hears that his main contractor has gone into administration, do you think he rushes around to the job to see that he's left it safe and secure? No. He or she thinks, how much do they owe me? When was my last invoice paid? They can see a hole in their funding right there. No way are they heading back to site. (The compound has probably been secured anyway by way of some hastily configured security arrangement. God knows who's paying them.)

In the case of the West Midlands Metropolitan Hospital, £204 million of the original £297 million budget had been spent by the time the main contractor went bust. £13 million was then spent just restarting onsite.

Much of the work had to be re-priced, and the cost of rectifying the existing work, plus the remaining work up to completion, was over £300 million. That's *on top of* what had already been spent.

The job is therefore going to cost almost double that of the original budget.

Who pays? You and me. You can't hold the new contractor to account. Let's just hope, this time around, they've got it right.

THE EDUCATION SECTOR

There is no doubt that education is of huge importance to the whole country, to individual families and to our political leaders. Everything is done with good intentions. Sometimes, though, those intentions can go to the wall.

In 1992, Tory Chancellor Norman Lamont introduced the PFI scheme as a way of introducing private funds into the public system to kickstart the economy and improve the country's infrastructure in a time of recession.

In the early Noughties, the scions of New Labour attempted to modernise Britain's places of learning, to turn them into veritable palaces. Money was to be no object. This was partly due to the fact that the costs of the government's wildly ambitious scheme would be spread across a generation or so, roughly 20 years.

One construction company at the forefront of delivery was Jarvis PLC. The company had grown out of the rail sector (and caused much controversy there for safety failings that contributed to some appalling rail disasters) before it sought to diversify into education.

They won a contract worth over £200 million to build or adapt eight new schools in the North West. They would then maintain them for the following two decades. This was a new dawn for the company – a chance to rebuild its brand and reinvigorate its tarnished image.

They needed to rapidly recruit the personnel needed to deliver the projects that they'd been awarded and had signed up to. I was lucky enough (or so I thought!) to secure an interview with them.

The project manager tasked with overseeing the entire scheme told me all about their plans, their pipeline of construction work, and their schedule for delivery.

The more he talked, the closer my jaw got to the floor. What he was proposing to do was virtually impossible. In my own life, I was looking for more control – a positive shift in my work/life balance. That was what I wanted, but the plan being outlined in front of me reflected anything but. I quickly realised that if I, and everyone else involved, committed to a seven-day working week, and then worked a 16-hour day attempting to fulfil our tasks, it would still have been unachievable.

And where do you go when a seven-day working week isn't enough? Do you go to eight days? Of course, you can't. Proof, therefore, that this really was an impossible ask.

Despite turning up late for the interview, they offered me the job.

I turned it down.

They then offered it to me again, this time with more money.

Again, I said no.

Yet more money was dangled in front of me should I choose to accept.

There was still steadfast resistance from me.

Even more money offered.

A final 'no' from me.

Five years later, as if to prove my point, I was brought in as part of a commercial and legal team to help extricate Jarvis from that same contract, which they had belatedly realised was unachievable. Well, I could have told them as much during that initial 45-minute interview.

Funnily enough, the firm that I chose to join at that time, instead of Jarvis, had recently lost a couple of their senior members of staff to that very company, lured away by the exciting opportunity and, no doubt, the higher wages that they were offering. Six months later, those two guys begged for their jobs back with their former (and my new) employer.

They came back with their tails between their legs. They had realised that no matter how hard they worked, they couldn't make a dent in the workload that Jarvis had signed up to.

Much of the work on those schools was scheduled to be carried out during the summer break. There was a deadline for completion at the start of September, for obvious reasons.

Sometimes, contractors just see the headline figure. They see an opportunity that's worth hundreds of millions of pounds. Surely anything is achievable for that amount of money? But break it down into bite-sized chunks and maybe the deal won't look so attractive. All that glitters is not gold, as the old saying goes. Where is Jarvis now? They went out of business in 2010.

Under the New Labour government, in the first decade of the 21st century, an expensive PFI scheme was created to build the schools of the future. That was pretty much the branding.

PFI stands for Private Finance Initiative, so although this was a Government scheme, and an honourable vision and dream, they didn't have the money to actually build the schools that these right-on politicians thought our children deserved.

Effectively, they were window shopping. They wanted the car, the house, or the new suit that they couldn't afford. But, as Francis Ford Coppola once said, 'Living within your means takes very little imagination', and New Labour were nothing if not imaginative.

So, what to do?

Well, here's how PFI worked. The contractors would bring money to the table, borrowing the cost of construction from the banks. The contractors would then be paid rent plus a 20-year maintenance contract for the same building, all at a level of remuneration in excess of the build cost and the interest on their borrowed cash. The local councils would then pay – over the 20 or so years – for the pleasure of occupying these new 'future' schools in their boroughs.

And the schools themselves were magnificent. They were veritable cathedrals of learning. Their scale was staggering. They were beautiful and brilliant. But they didn't come cheap. In 2019, *Schools Week* reported a JPI Media Investigations probe into how one school in Newcastle paid "£1,017 to move an LCD screen, while another paid £640 to remove a computer table and £560 to install a power socket." There are many more examples.

And here's a wrinkle. Guess who owns the schools and therefore controls them? Not the local authority. It is the contractors and their financial backers, (i.e., the banks or other funders who provided the initial private finance to build them).

Say the kids in Year 1 have an art class. They paint some lovely pictures as directed by their lovely teacher. Now, where are you going to put those pictures? *On the walls*, I hear you say.

Well, you can't. Because they're not the school's walls. The Blu Tack adhesive or drawing pins you're going to use to stick them up could be considered damage to someone else's property. They would scar the actual paintwork. That would mean increased maintenance and rectification costs. Not the school's walls. Not the school's building. Not the local council's either.

Private finance.

And some of these buildings, as fine as they were and are, were built in communities where pupil numbers were falling. So, they were mothballed and they became white elephants. Parklands School in Liverpool, despite closing for good in 2014, left an ongoing bill to pay in the region of £60 million at a rate of £4.3 million a year until 2028. Despite the school being built with the best of intentions, the council was burdened with 20 years' worth of loan repayments from the date of completion. That money has been legally and contractually ring-fenced. The council has had to pay, and still has to, for two decades.

That money will come from that local authority's budget, so when we wonder why there are fewer police patrolling the streets, or why there are more potholes in the roads, it's because a significant portion of the local authority budget has already been spent by previous administrations. Don't build what you can't afford.

COMMERCIAL, OFFICE, RETAIL, AND WAREHOUSING

The United Kingdom is still feeling the effects of the Covid-19 lockdown that undermined the British economy and saw almost two years of unprecedented flux to the way that we all live, shop, and work.

Over the past five years, up to 2023, commercial construction output has fallen by almost six percent. This is by far the largest reduction in activity across the entire building sector. In fact, it's the only area of the industry to record any sort of fall, and proof that we now have more office space than we need at the present time.

Working from home is – for many – the new norm. Certainly, it has become an expectation for many former full-time office workers. Some have hybrid working arrangements, part-time in the office, say two days a week, with the remainder based at your place of residence.

I can't see that changing any time soon. Try taking the new working arrangements away from people, many of whom, like myself, enjoy it. You'll see toys being ejected from prams across the nation. So ingrained will this new working pattern become, I believe, that I can't imagine much more than the odd statement-piece office building being built for corporate purposes in the next decade. Companies will either share or vacate their commercial premises.

Many of these properties might then be converted to residential use by their desperate owners. That's no bad thing. We need accommodation more than anything else right now.

The same challenges face our retail industry. That was true even before the pandemic hit. I'm not talking about supermarkets; they continue to post record profits, with sales increasing year on year. We all need to eat, so that makes supermarkets a pretty safe bet. But the high street? Dying a death, apparently.

Online is filling the vacuum. Making a fortune (not that the taxman seems to know about it). Out-of-town or edge-of-town retail parks are nothing new, but as the Amazon rainforest shrinks in the real world, its online namesake just gets bigger in the virtual version.

Warehousing, delivery drivers, the world at the customer's fingertips – that's the new normal for retail.

I'm lucky to live in an area with a thriving high street full of artisan enterprises. We have a butchers, a bakers, and a greengrocers, as well as the usual staples of Iceland, Home and Bargain, WH Smith, Tesco and Sainsbury's. There are also numerous cafés, bars, and restaurants. One of the boutique and bespoke stores close to my home is a rather nice men's clothing outfitter. I saw a T-shirt in the window recently and checked the price tag. £280! Needless to say, I didn't buy it.

I've not seen anyone go in that store. Not once.

The department stores that have shut their doors recently are too numerous to mention. Suffice it to say, now is probably not a good time to buy shares in retail. And what about some of those larger stores that lie vacant at the heart of many of our city centres? Well, some bright young things have had the idea of turning these empty retail and commercial office spaces into something else entirely. Apparently, they make excellent arenas for recreational use, which we're going to cover soon.

Oh, and by the way, the men's outfitters near my home are now into their second month of a 'one week only' sale. And that T-shirt I wanted? It's still there. And I'm still not buying.

SPORT, LEISURE, AND TOURISM

How's your work/life balance? What do you like to do in your spare time? That last question assumes that you have any. Some of you may not – you have busy lives – but a lot of us get to enjoy the occasional bit of downtime. We may choose to spend that in front of the television, or enjoy a night at the opera, theatre, or cinema, or take a trip to a sports or concert arena to watch a live event.

We may have a gym membership. We may even use it!

Most of what goes on in the world of sport, leisure, tourism, arts and culture, takes place within a venue. That venue had to be built. It needs to be maintained. These sectors contribute about £250 billion to the UK economy each year. That's more than ten percent of Britain's gross domestic product.

Some of these schemes are off the scale. The 'new' Wembley stadium (opened in 2007) cost over £1 billion to build, whilst a new home is one of the pressing issues for Chelsea FC's latest owners.

At the moment, despite having had one of the most successful teams in the country, Chelsea FC have one of the lowest ground capacities at just over 40,000. They barely scrape into the top ten Premiership grounds in terms of numbers.

As football used to be a working class sport, many grounds are situated deep in the hearts of their local communities, and Chelsea is no different. Except the surrounding houses sell for millions of pounds. Each. The idea of extending the existing footprint of the stadium, in its current locale, is simply not possible. So, the club will likely need to relocate and start from scratch, someplace else, in order to build a stadium of, say, 60,000 capacity.

Expected cost? At least £2 billion.

So, where will that money go, and how will it be spent? Well…

Consultants. That's the first significant chunk of outlay right there. Feasibility studies. Someone will need to identify suitable alternative plots of land that could accommodate a large-scale stadium. This is easier said than done in the Greater London area.

And it's not just a case of plonking a stadium into a plot of a similar size. There is all of the infrastructure required to go with it. How are people going to get to the ground? Is there a tube or train station nearby? Access roads? Parking?

Let's assume that the club's owners have found a suitable plot. They have heard and read the consultants' comments and reports, and they decide to go ahead with the deal. They'll need to pay a hefty whack to buy the land, inevitably involving a team of very well-paid lawyers.

A project management team will be appointed. There'll be an analysis of costs. Tender documents will need to be prepared. Main contractors will be sounded out. Quotes will be sought. Negotiations will ensue.

If you're lucky, within a year or two, you might see a spade in the ground.

The build costs alone will easily surpass the billion-pound mark. Then there's the fit-out. All those corporate spaces will need chairs and tables, carpets and furnishings. The same for the retail spaces.

Then there's the landscaping, hard and soft. Access stairs, steps, concourse areas.

Finally, access roads and the enabling works to facilitate their paths, which may involve buying up properties simply to demolish them as they are in the way. Then there's the parking, lighting, and services.

Manchester United might soon follow suit. Their ground is looking old and tired. Years of constant adaptations to the existing stadium have left the place groaning at the knees. The club can no longer work efficiently within the existing set-up and the current thinking is that it would be better to pack up shop and move sticks. Build anew.

I know that traffic around Old Trafford is a nightmare on match days, with the area virtually at a standstill for many miles around. Let's hope they get the infrastructure right next time and build the stadium somewhere accessible. It all forms part of a coherent plan.

As we are talking about leisure time, what about pubs? Since the start of the 21st century, Britain has lost a huge swathe of its public houses. From just over 60,000 such establishments in the year

2000, the number currently stands at just over 45,000. The demise has been placed at the door of high taxes on beer at the alehouse pump, and the low price of alcohol in the supermarkets.

So, what to do with all of those empty premises? Well, a lot of them are traditionally built bricks and mortar. They tend to be spacious, with generous proportions and a significant floor-to-ceiling height. They make for a desirable abode, and many have interesting features like grand fireplaces, cornices, corbelling and the like.

They are easily converted to living accommodations such as apartment blocks or homes of multiple occupation. There's life in these old pubs yet, with most being easily converted for residential purposes, and planning permission readily granted.

One bloke I know bought a country pub at auction at the height (or depth) of the last recession in 2009. He paid £180,000 for it. That was literally every penny that he could muster. It was going to be his final bid and, luckily for him, no one else in the room knew that, and they let him have it.

It will easily convert into two dwellings. It has a large car parking area in front, and to the rear it has the most amazing unspoilt views of the Peak District National Park. There are large gardens out back, benches and tables to sit at, and all encapsulated within a traditional stone wall no more than a metre high so as not to spoil the views.

You can't just build a new housing estate within the boundaries of a national park, therefore the vista is going to remain the same forever. He owns the property and enjoys the view.

A former pub can be a real residential winner.

And those large department stores mentioned earlier? The new use for some of them is as multipurpose activity centres. Go-kart racing around the fifth floor of your former Debenhams store. Just don't hit the plate glass windows! It's a long way down.

My point here is that we can develop new visions for existing buildings. That's the circle of life. That's the cycle of construction. Properties need to be dynamic, agile, and healthy, whichever sector they are in.

The tourism industry brings this section to its close. It's a sector that draws millions of visitors to our shores each year; those who want a dose of Premier League action, a blast from our music scene, our nightlife, or food at our Michelin-starred restaurants, but in large part it's our nation's heritage that attracts the tourists. They flock to our castles, museums, and visitor attractions. We even attract KGB assassins to our cathedral cities, such is their reputation overseas!

If those buildings are old then we need to maintain them because, if we just leave them to rot, maybe those visitors won't come. A symbiotic relationship between the built environment and the cultural lives of our population and our tourist industry is plain for all to see. These tourists come by land, sea, and air, and whether they are here to see Buckingham Palace, the Tower of London, or visit Loch Ness, they'll use Britain's trains and roads to get there, and stay in those places already built, whilst eating out and socialising in our bars and restaurants.

It's a relationship that will endure for decades and even centuries to come. It adds billions to the British economy each year, and pays wages, provides profits, and oils the wheels of industry. And the industry that benefits from it most of all? Well, that would be construction.

3. INFRASTRUCTURE

CIVIL ENGINEERING

What does civil engineering actually mean? I must confess, until I began writing this book, I only knew the broad brush definition. I just took it for granted that it meant the very big stuff in construction.

But, a little research later… and now I understand.

A long time ago – I'm talking centuries, and even millennia here – we never really had much in the way of big buildings in our towns and cities. And the world was a brutal place; it was all feudal fighting, a bit like *Game of Thrones*.

If you can imagine gazing upon our ancient landscape, and let's say that you happened to walk the length and breadth of Great Britain at that time, then (if you managed to survive) probably the biggest buildings you'd have come across would have been castles and forts – basically garrison towns and fortified buildings. Because, by all accounts, we were a very nasty lot in the old days.

So, the large-scale construction we had back then was effectively all military engineering, which meant huge stone edifices made with impenetrable materials quarried from rock faces that were built to repel all manner of assaults.

But then things calmed down. Law and order eventually prevailed, and people began to see the value in commerce and in manufacturing. Ergo, they needed factories and mills to produce their goods, and then roads and bridges, and canals and seaports to transport all of that merchandise.

We needed *civil* engineering, engineering for the benefit of civilians. Not just engineering of a military nature. And get it we did.

Civil engineering is basically construction writ large. It is a combination of brains and brawn. If construction is the building of things, then civil engineering is the building of *big* things.

Want to build a tunnel under a raging sea or a bridge across a large body of water? Call the civil engineers. Want to construct a nuclear power station or a brand-new national sporting stadium? You call the civil engineers.

At the time of writing, in 2024, many of the largest construction projects in the UK are civil engineering enterprises. These include the Capital's Crossrail system and its continued expansion and extension, the tunnelling and groundwork for the HS2 rail network (well, what's left of it), and the Hinckley C nuclear power plant. They each cost billions and billions of taxpayers' pounds, and the work that is involved will last for years and years.

Civil engineering is also responsible for giving us some of Great Britain's greatest pioneers and inventors. People like George Stephenson, remembered for his vast contribution to the railways. He not only built the bridges, but laid the track, and even designed the engines to run on them. Or Isambard Kingdom Brunel, who built so many of our docks, bridges, and tunnels. There's the visionary Joseph Bazalgette, responsible for London's original modern sewer system, which is now being revamped as part of Tideway, another huge civil engineering project to create a new super-sewer system in the capital. And what about Thomas Telford, who built harbours and tunnels, bridges and highways? So much and so many of the latter, in fact, that he became known as the *Colossus of Roads*!

We still owe a huge debt to these people today. We simply couldn't get around this country of ours, nor would we have done so for the last two centuries, without their incredible skills, their graft and their vision. Not to mention the wit and the will to bring their ideas to fruition.

But where are the modern equivalents? What are we building today? I don't see visionaries. I see government initiatives. We've done great things in the past; hopefully, we can do more of the same in the future.

GOVERNMENT PROJECTS

When we talk about the biggest construction or civil engineering projects, these are usually government-funded schemes.

Occasionally, a Shell or a BP or some other corporate entity might construct a new oil pipeline in the North Sea, which represents a massive undertaking for a private company. In general, though, the largest and the longest-lasting infrastructure projects tend to be financed straight out of central government.

Such schemes include HS2, the high-speed rail link that is meant to improve connectivity and journey times between the Southern regions of the UK and the Midlands.

The fact that our seat of government is based in London, which is the nation's capital and almost a country within its own right, means that focus and also financing has tended to lean (almost fall) towards the Greater London region. The North-South divide has even been confused with a North London/South London divide, such is some people's capital-centric thinking!

So, schemes such as HS2 (well, the original iteration) have been established to level the playing field. A little. Even that, though, has had to be watered down as the costs for the project have escalated with each passing week and month.

There's never a dull day in government. Someone, probably working in the department of fanciful ideas, gets to dream up the most extravagant and expensive schemes imaginable. Or unimaginable, for that matter.

How about a bridge to link Ireland with Scotland? Or a tunnel?

Both options were recently explored in a (bonkers) plan to try and span the Irish Sea (a distance of approximately 30 miles). The feasibility study alone cost more than £1m. The cost of building either a bridge or tunnel was estimated as recently as 2020 at no more than £20bn. When the idea was finally scrapped, early in 2022, the reason given was that the latest estimates put the cost of the tunnel at over £200bn and the cost of the bridge at over £350 billion. Well, that's inflation for you!

Part of the reason for abandoning the scheme, as well as the runaway finances, was the difficulty in pulling off what would undoubtedly have been a monster feat of engineering.

To give one example, there is a huge amount of unexploded ordinance, by which I mean bombs, dumped in underwater trenches since the end of World War 2, in roughly the same area as where the bridge or the tunnel would have been built.

When I say roughly, they only have a rough idea where these millions of tons of munitions actually are, as no *actual* records were kept about these dumping grounds. These explosive devices could be buried anywhere within a multi-mile radius.

Obviously, the bridge or tunnel would seek to cross the divide between Scotland and Ireland at the shortest possible point. Unfortunately, the shortest point, in this case, would appear to go right through the world's most crowded minefield! Or thereabouts.

Of course, not every government scheme fails to come to fruition or is in danger of blowing everyone to kingdom come. In fact, every scheme of national importance is appraised by the government through a recently formed department known as the Infrastructure Projects Authority, where the merits of each proposal are measured against certain criteria, including the likelihood that the idea can be brought in on time and on budget.

Schemes such as the Scotland/Ireland bridge or tunnel link were ultimately rejected as the overall cost was considered to far outweigh any possible future benefit. And one reason for that false or inaccurate early cost forecast was – surprise, surprise – the political nature of the very scheme being proposed.

Some ambitious person in Westminster tries to sell a project to their political masters, who then run with it and try to sell it to the British public. Short-term politics makes for unrealistic budgets and – should the project be green-lit – the taxpayer will ultimately have to pay the true cost.

Fancy a new Scottish parliament building with an estimated bill of between £10 and £40 million? Actual cost of Holyrood, as nice as it is, at £414 million… in other words, add a nought to the higher end of those overly optimistic figures. Or HS2 at almost double the predicted cost. The Millenium Dome coming in at about the same, (i.e., twice what they initially told us and sold to us).

So, the next time a large government construction or infrastructure scheme crops up on the evening news, just remember to add an extra nought on the end when they eventually get around to calculating the budget.

UTILITIES & POWER

What would you do without your mobile phone? How would you live or be able to work without the internet? When was the last time, if ever, that you experienced a power cut?

The truth is that we have all become a little pampered in our daily lives. No one keeps an emergency supply of candles any more for when the lights don't come on. We just expect all of this stuff to be there, and for it to work, whenever we need it, which is all of the time.

A lot of the companies that provide all of these things are now privately owned. By which, I mean that they're in corporate hands, rather than being government-owned. 'Selling off the family silver,' that was the chant of those opposed to the privatisation of our services when these things were happening during the last quarter of the 20th century.

For better or ill, and regardless of who now owns these companies, we would all struggle to live without their products. We all need gas, water, electricity, and data connections.

Much of this connectivity now takes place underground. We have pipework, ducting, and cabling to ferry these services from their point of origin to our homes. Just be careful where you're digging, as roughly two construction workers a week suffer life-changing injuries as a result of striking live electricity cables and the like as they go about their work.

We have nuclear power stations. We used to have coal-fired power stations, and we are now moving towards greener technology, such as wind farms and solar panels. In 2021, the UK government introduced its Net Zero Strategy to reduce greenhouse gas emissions to nil by 2050. Utilities and Power lie at the forefront of this lofty endeavour.

Within a decade or two, it is also expected that petrol and diesel cars will be obsolete. Our vehicles will all be electric or powered by hydrogen cells or some other invention.

Charging points are now popping up everywhere. Aside from at people's homes, you'll find them at almost every petrol station and motorway service station. Even restaurants and fast food outlets are installing them on their premises. You can recharge your car battery as you grab a bite to eat.

United Utilities and Thames Water are two of our biggest fresh water and water treatment companies. It's no secret that we lose billions upon billions of litres of much-needed water each year as it seeps into the ground through broken and ageing, rusting pipework.

I once heard a comedian exclaim on stage that human beings are the only animal that urinates in their own water supply. That's sad but probably true.

And speaking of comedians, some of our finest – including Steve Coogan, Paul Whitehouse, and Lee Mack – recently attended a rally in Windermere to protest against United Utilities and their appalling record of releasing raw sewage into the Lake District beauty spot and its environs. In 2022 alone, there were 5,900 instances of untreated sewage being poured straight into the lake's catchment area by just this one company. They then had the gall to announce a windfall of £300 million to be returned to their shareholders. This is obviously money that could and should have been spent elsewhere, for example in cleaning up their own (and our) mess, as is their remit.

We release raw sewage into our lakes and rivers, but when we send the result of our ablutions *offshore*, we don't send it out far enough. We don't even give it a chance to disappear, and it comes straight back at us on the next incoming tide.

So, we need to fix those leaking pipes. We need to explore less harmful forms of power. We need to capture the rainfall and learn how to desalinate water from the oceans cheaply.

Construction plays a huge part in the services industry, and there is a growing synergy between the building and service industries. They keep robbing our personnel, for one thing. Electrical companies, IT and data suppliers, and the like, are forever advertising for engineers, contracts managers, quantity surveyors, and the like. It's clear that this is a growth industry, providing the energy and the vital services that we all need, but construction needs all the trained staff it can get, so I suggest the utilities and power companies should train up their own. They certainly seem to make enough money. Why don't they recruit the next generation of apprentices and go from there?

Going forward, countries, including the UK, need to be responsible for sourcing their own energy. We basically have to become self-sufficient. If not, then geopolitics or world events, such as the Russia-Ukraine war, can disrupt things enormously.

Ever heard about the company in Germany that won a contract to make uniforms for the Allied armies in the Second World War? It beggars belief, but it does go on.

So, while people across Europe reviled Russia for its aggression in 2022 against Ukraine, they were still reliant on Putin's gas supply to fuel their ovens and radiators. We had all basically done a deal with the devil. Our opprobrium could only go so far, and we had to say, 'We don't like you, and disagree with what you are doing, but can we please buy some more of your energy? My boiler runs on gas and I don't want to be cold.'

The solution to a potential energy crisis? It's quite simple. Grow your own. Find your own. Build your own.

Whether that be nuclear, gas, wind, solar, or a hydro-electric dam, we need to generate our own power sources. That way, we can't be held to ransom by countries that we'd rather not do business with, and we can't then be cut off if we fall out with our geopolitical neighbours.

Build reservoirs. Capture energy. Make provisions for the future.

TRANSPORT

The roads and the railways, the bridges and the tunnels, the ports and the airports that facilitate the movement of people and goods from A to B. All of these infrastructure items are built by the construction industry, and often at great expense.

The Channel Tunnel Rail Link, also referred to as HS1 (the initials standing for High Speed), cost £5.7 billion for a 67-mile length of track that opened back in 2007. That's nearly £90m per mile, or £60k a metre!

The budget for HS2 to improve services from London to the north of the country had risen from £55 billion in 2015 to over £100 billion in 2023 before it was scaled back. Before the decision was made, half the scheme had been 'red-flagged' as unachievable. In January 2024, the boss of the company behind HS2, Sir Jonathan Thompson, said the London to Birmingham section alone could cost more than £65bn.

Meanwhile, Crossrail in the Capital, now renamed the Elizabeth Line, opened in 2022, just in time for the (now, sadly) former Monarch's 70-year jubilee. It came in three years late and almost 20% over budget at close to £20bn. Not red-flagged that one, of course. London-centric, a lot of this stuff.

The year 2025 will see the 200th anniversary of Britain's (and indeed the world's) first public passenger rail service. It is a noteworthy achievement and something which the country can be proud. I hope we do George Stephenson and those early railway pioneers justice when we come to acknowledge their vast contribution to British transport.

At present, almost half of the UK's annual £50 billion spend on transport goes to the railways. Another 25% goes to the roads. That was the split as of 2021.

Cycling is also a major focus for the government. I wholeheartedly believe in cycling as an important part of the mix, too, and I am a big advocate of cycle lanes. From the age of ten to 15, I spent every summer in Holland, staying with my aunt and uncle and cousins. We'd go camping near the German border, where we would pull a trolley into the site, filled with tent, dinghy, hampers, and idle infants.

Barely ten feet away would be a noisy and busy dual carriageway, but between us and that road there'd be a reassuring border of tall trees and also a substantial concrete kerb. Before that, there was a cycle path and then another kerb. And only then the pedestrian path. All properly differentiated. All practically safe.

In the Netherlands, they cycle around 15 billion miles a year across the population. Fatalities are about 200 per annum.

In Britain, we cycle around 3 billion miles per year. Fatalities in 2020 were 141. Our roads are, therefore, 400% more dangerous for cyclists in the UK than they are in Holland. Why's that?

Well, I would say that in the Netherlands, they have proper cycle lanes. In the UK, we paint lines in the road and call them cycle lanes. They're not. They are still main roads; vehicular roads. And that white line on the tarmac won't save your cycling ass one bit!

The city of Manchester has got it right, though. In 2017, they appointed Chris Boardman – the Olympic gold medal-winning cyclist and cycling advocate – to advise on their cycling and pedestrian routes, and spent over £13 million delivering his vision. It made Manchester one of the most cycling-friendly cities in the world.

Boardman is someone who knows what they are talking about, knows what they are doing, and has a passion for the subject. During his tenure as Commissioner for Cycling and Walking, he produced a 15-step report entitled *Made to Move*. The overall concept was designed to make walking the natural choice for as many short trips as possible, and to double and then double again the amount of cycling taking place across the region. The aim was to create a genuine culture of cycling and walking. This wasn't lip service. This was real change.

In 2021, Chris was appointed Transport Commissioner for the whole of Greater Manchester, tasked to deliver a fully integrated transport system for the entire city region. Then, in 2022, he was poached by the British government to roll out his exemplary vision on a national basis. I say *Amen* to that.

The Active Travel England scheme, of which Boardman is the Chief and Chair, will now have an automatic place at the decision-making table for every major transport project in England, advising on how the needs of cyclists and pedestrians can be incorporated into each venture.

Personally, I think that every major road or transport route presented for planning approval should incorporate a cycle *and* pedestrian route along the same corridor. Should the proposers wish to opt out, they would need to provide significant reasons why that particular scheme can't incorporate such an addition. They should have to prove *why* such a thing is practically impossible to achieve. And there should be no easy get-out.

The major expenditure to construct any new road is in the land acquisition, the professional and planning fees, the levelling works down to the formation level, and then in the building of the road itself. The extra cost to install a secondary kerb, say a third of a metre high, to separate a cycle lane from the main highway, would be marginal. It might add between one and five percent to the overall budget.

Let's legislate this. Let's make it compulsory. Unless the developers and decision-makers have a very good reason (and in most cases, I can guarantee they won't have) for making their scheme 'road-only', then let's make all our new roads and planned routes available to everyone.

DEFENCE

After transport, the defence of our nation is where we spend much of the remainder of our infrastructure money.

We spend 50% of our infrastructure budget on transport, and about 30% of it on defence.

The amount spent on defence includes the cost of missiles and machinery, of course, but a significant proportion is also spent on housing and accommodating the needs of our service personnel.

In 1996, the MOD scandalously sold off 57,400 of its properties to a private company for £1.7bn. It then agreed to lease them back for the next 200 years. At present, they are paying rent of £180m a year on those dwellings. Annual maintenance and refurbishment costs are another £140m annually.

The average cost of a house in the UK in 1996 was £55k. In 2023, it is £275k, a fivefold increase. Likewise, the current value of the MOD's portfolio of *sold* properties is now £7.6bn (again, a fivefold increase). But, at the current rate of rent returned to the buyer (Addington Homes), the money the MOD received will only cover a decade's rent. That still leaves 160 years of signed-on-the-dotted-line rent to pay for, and that's after the new owners have already had 30 years' worth of rent from these homes!

Does that sound like good business to you?

I guess not. And nor does it to the British government. Plans are now underway to render the original sale null and void. It was a really bad deal for the country and the taxpayer.

Even if the government is successful, I can only imagine that extricating ourselves from our contractual obligations will not come cheap. The words 'loss of profit' come to mind, and will surely be mentioned by the new owners' barristers. Watch this space.

Still, if the buyer wants to get stubborn and refuses to accept the government's offer, I suppose we could always send in the tanks…

Despite that mammoth property deal, the Ministry of Defence still has a large number of dwellings in its portfolio. About 100,000 individual residences. The MOD is required to keep about ten percent of these spaces vacant at any one time, available to military personnel who are being transferred at short notice. At present, the percentage of vacant properties is closer to the 20% mark, which has led to calls that some of these properties be released onto the housing market to help ease the shortage of affordable homes nationwide or, more poignantly, to house veterans of our armed forces who might be living on the streets.

The MOD remains unmoved on this matter for the moment. You never know, professional troop numbers might rise again from their current all-time low, and these empty homes might be brought back into use after all. Until then, they will need to be properly maintained. That's where the local or national maintenance provider comes in.

Nice work if you can get it. I've worked on a variety of building projects on a number of different military bases.

There's obviously a high level of security involved to get you and the tradesmen onto site, after which it's just a regular construction job. But not always.

I once went to an army camp in North Yorkshire. A company we had a relationship with had the building and maintenance contract for that military base and – in essence – they were little more than a project management firm, as they employed no skilled tradespeople of their own. How they had then won the contract, I have no idea. Good branding? Good salesmanship? A little baksheesh? Anyway, all of the *actual* work was subcontracted out to us.

We'd previously put some new doors onto an electricity substation lying just off the runway of their main airfield. Our client subsequently phoned to say that we'd fitted an internal-quality door to the exterior of the building. In the six months since we'd fitted the wrong type of door, it had rotted, warped, and fallen foul of the weather. We needed to go back and replace the damaged one with a robust external-quality door, which is what we should have done in the first place.

I arrived onsite at the military base, went through security, and then waited for our liaison to come and show me the problem, so that I could agree that we had made a mistake initially and then put it right.

It was a case of measure up, order the new exterior door, and send a joiner to the site to fit it.

'Follow me,' the project manager bloke said, before he jumped into his impressive pick-up truck complete with 'flashing light' roof bar. I climbed back into my own little motor to be escorted to the substation, which was a mile or so away across the base, out to the airfield.

I followed him until he eventually came to a halt. We climbed out of our vehicles and we walked down a path to the concrete bunker that housed the electricity generator. I then took one look at the warped and damaged door and immediately agreed to replace it.

Easy-peasy.

I took out a tape measure and measured the frame of the door so that I could be sure to get one the right size. I then walked back up the path and got in my car. And waited. And waited.

My client didn't seem in any rush to get back into his leading vehicle and escort me offsite. Indeed, I couldn't see any sign of him. Maybe we were finished? Perhaps I was no longer needed?

After a few minutes of this, I thought, 'I'll just head off'.

Now, which way had I come?

There's the control tower there, I determined. Now, was it left at the control tower or was it right at the tower? I honestly couldn't remember.

I started the car, drove off, and took a left.

This is a nice road, I thought. Smooth surface. Very even. Very long. Very straight.

Looks more like a runway.

Shit!

It *is* a runway! It's the actual bloody runway!

I veered off sharply onto a tarmacked car park area, and from there, motored through the camp's service roads until I reached the main gate. I hastily signed out at security, then headed for the motorway and home.

Five minutes later, my phone rang.

I answered it. It was my client. He was, perhaps unsurprisingly, not very happy.

'I've just had a right bollocking because of you,' he spluttered. 'They had to divert a plane in the air that was on its final approach to landing!'

And a military airplane at that. Whoever said construction was dull!

But that was just the story of a little job. I've also been fortunate enough to be part of one of the biggest defence spending building projects of the past decade, a £168 million refurb of RAF Lyneham, now renamed MOD Lyneham, that began in February 2014 after planning permission was granted by Wiltshire Council in 2013.

The upgrade/overhaul of the existing site was intended to take this large military airbase and turn it into one giant new training headquarters for Great Britain's Army, Navy, and Royal Air Force. Here, they'd all share a home for their new recruits, and little expense would be spared to make this a camp that all of our armed forces could be proud of.

The RAF decamped from the base while the building work took place, while we – a veritable army of builders – constructed everything the military could possibly need: new accommodation blocks, new medical facilities, a new sports centre, bars, restaurant, tank repair workshops, offices and stores.

And we finished on time and on budget. There aren't too many such examples around in major public works.

You could almost have called it a triumph except, halfway through construction, when the three branches of the Forces were asked about ownership, the unthinkable happened. Two of the three expected tenants were decidedly unimpressed. It was as if they had never been consulted in the first place.

The RAF, for one, said, 'As you know, we decamped to Brize Norton while the building work was being carried out, and we've decided that we rather like it here so we're going to stay put.' They weren't going to return to their newly refurbished facility, and their new home in Oxfordshire is now Britain's biggest RAF base.

The Navy then gave their verdict. 'It's miles away from the sea,' they said. 'We were never going to come.'

I mean, did anyone ever ask these people in the first place? Has anyone in the MOD ever heard of market research?

It might have been a good idea for someone in government to have got the names of the Chief Admiral, Chief Air Marshal, and the Chief of Staff of the British army and to have asked them the question, 'If we build you all a £168 million joint HQ in the Wiltshire countryside, would that be of interest to you?'

No. Didn't happen. So, it has now become the sole preserve of the regular army. Theirs to rattle around in, such have our troop numbers fallen in the last 50 years.

This was a great construction project, but only one of the three envisaged clients were actually interested. Wowsers.

Today, all large public schemes are appraised by the government's Infrastructure Projects Initiative team. They employ 17 commercial managers to do the initial number-crunching for planned major works. They present the headline figures that then get bandied about in parliament and on the news. These are the figures that we can usually add a nought to (at the end) when and if these schemes actually come to fruition.

They might be better called 'The department of wishful thinking and make-believe'. At least when they do spend their (by which I mean 'our') money, a lot of it ends up in the building industry.

Infrastructure. It's big. It's occasionally wasteful. But it's still construction.

4. MEET THE CLIENTS

THE PRIVATE DEVELOPER

Approximately 40% of UK construction is carried out by private developers. The range of what these companies and individuals do covers every aspect of the building industry. This is not just someone constructing a new house on a vacant plot, or knocking down an existing (and often historic or interesting) property in order to throw up a block of flats with a much greater footprint (e.g., eight floors instead of the current four). No, what we are talking about is the entrepreneur at work here. A risk-taker. Someone who has spotted a gap in the market and is prepared to stake, if not everything, then at least a significant amount to realise a vision that they believe in.

You've got to admire them for that. For the vision part, at least. The risk element is usually, almost universally, mitigated at the initial brain-storming and planning stage.

No matter how much cash these entrepreneurs have in the bank, it seems that they all learned something very important on their climb up the financial and property ladder.

Never use your own money.

Having worked so hard (presumably) to become rich, they aren't going to gamble it away with a roll of the dice on a construction project. Even if the rewards appear astronomical, the element of risk seems to be enough to bring realism to their dreams. Only once the cost of borrowing the money to fund the scheme has been factored into the overall budget, then (and only then) will a project be deemed remotely feasible.

I was briefly employed by a private developer who specialised in housing schemes in some of the most affluent parts of the south of Britain. They would scour the land banks for suitably sized plots, capable of spawning between 20 and 100 new homes.

Before they would even consider entering a bid on the opportunity, their thinking – and their budget cost exercise – went something like this.

Cost to buy the land? Say £5 million.

Suitable for how many properties? Say 100.

Average footprint of each property? Say 2,000 square feet.

Overall footprint of the scheme? Equal to 100 properties multiplied by 2,000 feet. A total of 200,000 square feet.

Build cost per square foot (based on the current industry average) is £200.

Total build cost is therefore 200,000 x £200 = £40 million.

The total scheme cost is therefore £5 million for the land plus £40 million for the build.

Add a further ten percent for landscaping and fees.

We now have an outlay of £49.5 million.

The cost of borrowing that money is, say, five percent. Sales fees for estate agents are another five percent.

That's a grand total of £55 million. It soon adds up, doesn't it?

In return, the developer would have 100 brand-new homes to sell. On the south coast. In a very nice area.

The selling price is expected to be £600K to £650K. An average of £625K.

That's a project outlay of £55 million with a return of 100 units multiplied by £625K, which comes to £62.5 million. That's an overall expected profit of £7.5 million.

That's just shy of 15%.

Given that the developer is using someone else's money, and has factored in the cost of borrowing that money, that doesn't seem such a bad return.

Say there are five investors in this private development syndicate and say it takes three years from start to finish to complete the scheme. That £7.5 million works out at £1.5 million per investor. That's half a million, per year, that each of them have earned for the three-year duration of the scheme. That's not bad work if you can get it.

And for the developer? Everything on someone else's dollar. Arses covered, although – of course – it is not risk-free.

Take the model village of Bournville in southwest Birmingham, for example, built by the chocolate dynasty – the Cadbury family. Old Mr Cadbury had a very good recipe for chocolate. He thought that the populace had a sweet enough tooth for his product and he needed a factory in which to make it.

Then, having built the factory, he constructed an entire village in which to house his workers. (Cadbury was a Quaker and wanted proper housing for his workforce.) Having provided his employees with both a job and a decent home, he could be sure of one thing: a loyal workforce. That's some clever thinking right there. And the chocolate's not bad either!

The Cadbury family are just one example of the entrepreneur at work.

There is a beautiful block of flats in the south coast resort of Bournemouth called the San Remo Towers. It is on a hill on a road that leads down to the beach and one of the town's piers. It is a fabulous building. I fell in love with it the first time I saw it.

It is Moorish in design, and more-ish on the eye. And its location is simply divine.

It is five storeys high, and about 100 metres long by 60 metres wide. There are over 150 apartments in the development. Want to buy one? Well, the good news is, you can. But, when these properties were first built (in the mid-1930s) by a private developer, none of the flats were for sale.

You could rent one, and even then you had to be appropriately wealthy and you had to provide suitable references in order to be able to do so. The thing is, these were very desirable properties, in an extremely desirable location. This became a sought after address because of the limited number of available plots and the restrictions on tenancy.

Now, I'm sure that many of the tenants who came on board thought that renting was just folly. It's dead money, right? We all know that. But, in the right circumstances, it is possible to convince even the wealthy and no doubt worldly-wise amongst us that something is worth having, if only temporarily.

Who were the smart ones in all of this? The private developers. The unique visionaries. The people with the plan. That's one they definitely got right.

Much later on, the premises at San Remo Towers were sold off as a whole lot. Lock, stock, and barrel.

The new owners then divvied out the apartments. Sold them all off individually. The building is still there. Still on the same spot. And it still holds that iconic glamour. Only the original business model is dead. For me, that's a shame. I think it worked wonders.

The road of the private developer is never paved entirely with gold, however.

I know these three lads. Jack The Lads, you might call them.

They reached a certain age in life (let's call it maturity), and with families to feed, they put behind them their childish (and slightly naughty) ways and decided to try and earn honest livings for themselves and their partners and dependents. Nothing wrong with that.

They partnered up with some experienced developers who were probably a bit further down the road of respectability than they were, and delivered a housing project. The scheme went well and a profit was made.

They wanted to go again, so more land was acquired and more building work took place. More profit ensued from their endeavours, and it looked like these boys were onto a winner.

For their next venture, a large derelict property was found, with the potential for even greater development on an adjacent plot. Money was raised from external investors from across the globe. About £10 million to be exact. Not small change.

Building work commenced and it seemed to be going well. Except these lads weren't really builders. They had a bit of experience but no expertise; they were winging it.

Although their intentions were honourable and their ambitions were admirable, they hadn't drilled down into the level of expertise and knowledge required for a project of this size. A visit from the building inspector and the local fire officer – just before they were to welcome their very first tenants into phase one of the project – led to the building being abandoned, deposits returned, and – ultimately – the development company being put into administration.

So, what had happened?

Simple really. They hadn't resourced the scheme with qualified construction professionals; the architects, engineers, and all manner of experts who could (and would!) have identified and rectified the glaring holes in the construction of these buildings.

They thought they were capable of managing a project of this size, when really they weren't up to the task. You need to know your limitations and put the right people in place, as and where you need them. So they saved a bit of money by trying to do everything themselves, these boys, but they had bitten off more than they could chew.

As I write, the buildings themselves are still incomplete and the whole project was recently put up for sale by the administrators. At auction, the development was picked up for a knockdown fee, presumably by someone with the nous to build it properly, whilst the budding developers are back at square one.

In years to come, people will drive by or walk past those occupied buildings, never knowing the dreams that are buried in the foundations. The dreams of those daring people, part of that insightful and occasionally reckless private development breed.

My point is, construction isn't easy. It's no different to car engineering or brain surgery, (i.e., best left to the experts). And though these boys had done an excellent job of *selling* their dream to investors, they couldn't actually build it themselves. And instead of appointing people that could, they tried to wing it and came unstuck.

I know a bloke who runs a successful painting and decorating business. He makes his bread-and-butter in the construction industry working for main contractors, and has a few sidelines. He's an entrepreneur, but a practical one.

He spots gaps in the market, say for student accommodation in a bustling city. He builds said accommodation. Except he doesn't actually build it. He has the idea, he secures financing for the whole enterprise, but he lets a builder build it. Sounds simple, doesn't it? It's common sense. The head ruling the heart.

Private developers come in all shapes and sizes. What unites them is vision and graft. That, and the financing that helps make their dreams a reality. Even if the money behind it is often someone else's.

But leave the actual building to the experts. That's what the smart folks do.

THE HOMEOWNER

The individual homeowner is one of the most important clients for construction work.

The recent lockdown, and the subsequent increase in the number of people working from home, as well as the incumbent uncertainty around overseas travel and foreign holidays, have all served to make people look closely at the four walls in which they live.

The result? Many people aren't happy with what they've got, and home improvements become the order of the day.

This is generally the preserve of the little guys in the construction industry. Small firms; a few tradesmen banding together and each doing a bit. For larger jobs, say upwards of £250k in value, a medium-sized contractor might answer the call, but it's mostly those little guys who come in to do your kitchen, bathroom, or extension to the rear of your property.

Right now, this is a golden age for small builders. They are very much in demand, although I personally think that their time in the spotlight is richly deserved. This means these guys (and, of course, gals and others) can now pick and choose. They might be able to fit your project in at some point, and if not this year, then hopefully the next. Oh, and the price will be this.

Not interested? No problem. They're busy anyway. Enjoy your four walls as they currently stand.

Unfortunately, a lot of clients don't know the difference between a good builder, or how to find one, and one who simply talks the talk. Many are the tales of woe caused by the latter. By the same token, small builders don't know how to tell if a homeowner will prove to be a good client or a right pain in the backside. Hopefully, I can help you here.

First up, some words of wisdom for the builders. When you work for a homeowner, it's important to remember a number of things.

Firstly, you are working in the client's home. As such, they'll be all over you like a hawk. This isn't a building site; it is somebody's property. It's not their investment; it's the place where they lay their heads every night. As such, they'll demand almost perfection. Not that they'll want to pay you for it, but it's probably wise to factor in a bit of nuisance money when you come to quote for the work.

Also, the homeowning client is unlikely to be a construction professional. They might not know what they want, or know how to articulate what they think they want, so be absolutely clear on the scope of works, and the specification (i.e., the quality of that work), before you even begin.

My advice to the builder is to take the lead. Use your experience to guide the client as to what is achievable, practical, and will fulfil their project aims. Agree the scope, the specification, and the price in advance. Then, ask them nicely to please step aside and leave you and your tradespeople to get on with it.

You can't have them looking over your shoulder or whispering in your ear every hour of every day, so schedule a weekly update meeting. Set it for one o'clock on a Friday afternoon, and during the meeting, offer the construction plan for the following week, so they know what to expect and what's going to happen next.

Then get them to give you ten grand (or 20) to pay for the work done that week.

That way, happiness lies.

I've seen clients change their minds umpteen times while the job is onsite. They're not sure if they want it done this way or that way. The builder is standing around, driven to distraction, wondering which way to turn. All the while, time is dragging on. Essentially, the client has a full-time handyman on call. Except they won't want to pay for it.

The home renovation market in the UK is currently worth over £100 billion a year. More than 50 percent of British homeowners recently surveyed said that they were planning to carry out home improvements during or following the pandemic period. Whilst some of this will be DIY, there is still a vast tranche of work for Britain's army of small builders. The individual homeowner is of massive importance to the daily activity of the UK construction industry.

LOCAL AUTHORITIES

A lot of building companies make a tidy living by being on the books of their local authority. There are a number of hoops to be jumped through to get on the approved contractors' list, but once there, these companies can then usually breathe a huge sigh of relief.

Local authorities are responsible for the social fabric of their regions. They have statutory obligations to maintain a certain standard of civic support, for which they charge the local residents a council tax, and they also get an annual budget from central government to help fund their activities.

The thing is, these local councils don't like to get their own hands dirty when it comes to construction. They might employ their own surveyors to manage and monitor the work, but the actual physical work – building-wise – is, for the most part, done by these lucky others.

At one time, if a council tenant had a broken window or a problem with the front door, they would call up 'the corporation' and a job would be logged. Eventually (and this could have followed a wait of weeks or months), someone would come out to carry out the repair.

This is no longer the case. When Margaret Thatcher was prime minister, she began a major re-modelling of British society. She was keen to promote enterprise in every corner of the economy, including the staid world of local government.

Previously, each council had employed its own skilled workforce of tradesmen. But now, competitive tendering was the key. Value for money was (meant) to be had, which meant that those secure council jobs for local tradespeople were no longer there.

Look at the cleaning of hospitals. Each ward used to have a matron. She (it was almost always a woman) would take great care of the environment she controlled, and patients knew that they were in a good place to tackle the illness that had taken them there. They would probably emerge in better health than when they went in!

But where does that attention to detail appear on the balance sheet from an accountant's point of view? Nowhere. Save a few quid. Outsource the work. Get cheaper personnel. They won't do half as good a job, but you saved ten percent on the cost. Makes sense, doesn't it?

Contractors value the continuity of work that a local council contract can bring, but it does not produce much of an actual saving for the consumer, meaning the taxpayer. Local authority work accounts for about ten percent of construction work across the country. It is a valued pipeline for builders, but I don't believe that it has produced much benefit to the people that ultimately pick up the bill, which is everyone reading this book.

Maybe the £4 billion of levelling-up funding, for which these local authorities can now apply, will reap some rewards for our more impoverished communities and, again, our building firms can do well out of this work, as long as they're on the approved contractors' list.

THE CLIENT'S REPRESENTATIVE

Not every client feels suitably qualified to engage and interact with their builder. In that instance, they usually reach out for a helping hand.

Say you're a local council, a school, a manufacturing plant, or some entity outside the construction world, but you need some building work done. Well, why not appoint someone who can represent your interests? Makes sense, doesn't it?

If you watch *Homes Under The Hammer* on television (I call it Homework Under The Hammer!), you might hear an enthusiastic amateur saying that they are going to 'project manage' their envisaged scheme themselves.

Good luck to them, I say, and I honestly mean it. But, while this might occasionally turn out okay – as conducted by someone who wants to learn on the job, is willing to get their hands dirty, and has a degree of common sense that enables them to achieve good results – it is often just a cost-saving exercise. Why pay someone to manage the job when I can do it myself? The work then ends up costing *more* than it would have done if they'd employed a professional to act on their behalf.

I recently took a phone call from a distressed homeowner who had some building work done on their house. They had no one in a professional capacity acting on their behalf, representing their interests. They were, to all intents and purposes, at the mercy of their builder.

Their simple wraparound extension had become a job that had no end. They were pouring money into the scheme while their chosen builder, in their view, simply led them up the garden path (or around the extension). They were now in the process of suing their builder. And he was counter-suing them.

The people at the end of the phone wanted me to act as their expert witness, and to comment on the work that had been done to date: what I thought it should have cost, how long I thought it should have taken, and all of that.

I went with my initial gut reaction and said that I really didn't want to get involved. Why would I? The project was already in a mess, and I knew neither of the parties involved.

I did read through some of the correspondence they sent me. If I thought I could help, I certainly would have, but the exchanges were a classic tale – a tale that plays out across the country regularly. In a nutshell, the homeowner said that the builder was taking too long. In reply (and this was all in the email trail), the builder said that the homeowners seemed to change their minds on a daily basis as to what they wanted done. He was chasing his tail, and they were paying for the privilege of watching him do it.

It was a real mess.

When I explained to the homeowners that I could offer them little other than an overall independent review of the scheme, without wishing to fall out with anyone, the penny appeared to drop for the client.

'We should have got you on board earlier,' said the lady of the house.

Yes, I replied. The builder needed clear direction, whilst the client needed an actual quote based on a defined scope of works. Variations and fluctuations can be monitored and managed as the work progresses, but you can't write a blank cheque. Nor can you really blame the builder for treating it as a casual day job. He basically hung around until the client decided what they wanted doing. The fact that the homeowners kept moving the goalposts was hardly the contractor's fault.

Anyway, this is why a lot of people, companies, councils, and private enterprises turn to an experienced hand to protect their interests. That's the role of a client's representative – and it could be fulfilled by an architect, project or property manager, or a surveyor of sorts.

Let's use an architect for our example. Say a client has a vision and the necessary finances to realise their ambition, an architect is appointed and draws up the plans. They can then reach out to three or four contractors to obtain quotes for the work before advising the client as to which of those quotes to accept. It might not be the lowest price received, but then there also has to be a very good reason for not going that way. It could be availability. The builders have a full order book and can't start for six months.

Once the discussion has been had, and the choice of the contractor made, the architect can then draw up the contract and oversee the work from that point on. They will then evaluate the completed works in terms of quality and quantity, and can approve the contractor's invoices as and when they are presented, whether that is on a weekly, fortnightly, or monthly basis.

Yes, there's a fee involved for this governance and support. Maybe it's 10 or 15% of the total build cost, but it's money that should be considered well spent. It saves the client the headache of doing their own project management, for which they may be entirely unsuited. It also ensures that they get what they are paying for.

Oh, and if it does go awry, the client can still (though please don't!) sue the architect against their or their company's professional indemnity insurance.

The client's representative does not have to be the architect, of course. You can leave the architect to do the design, and then employ a quantity surveyor to act on your behalf. The latter is more money-orientated, and you can (usually) trust such a person to keep a tight hold of the purse strings for your project. Whichever professional you choose to represent you, the fees involved, as a percentage of the contract value, are all going to be around that same 10 to 15% margin. Nowadays, there is a burgeoning market in professional personnel simply called 'client's representatives'. They do what they say on the tin and are essentially management consultants with professional construction backgrounds.

They will (or should) tick every box and fulfil every task for you along the way. They'll be your interpreter, your sounding board, and also your shoulder to cry on. They might be your punching bag, too.

Client's representatives can help bring order to potential miscommunication chaos. Even the smallest job can benefit from having a knowledgeable person on the side of the client. An intermediary between the parties, no doubt erring on the side of the client (who is paying their wages) but also listening to the contractors' concerns. Indeed, the builder should embrace the presence of a fellow professional on the project, for they might also need that interpreter. Or a shoulder to cry on, too!

5. MEET THE CONTRACTORS

THE LITTLE GUYS

Many of the people employed in the construction industry are the little guys and gals who operate as sole traders or within a very small team. Think of all those white vans that you see driving around on Britain's roads. Often badly.

Mick the Gutter Man. J and B Roofing. CJ Electrics (Domestic), etc. The list is pretty much endless. These people are the bread and butter of the building industry.

There are some five million small businesses in Britain. This is anyone employing between one and 50 workers. Approximately 20% of all these small businesses are construction-related.

In turn, of the three million people employed in some capacity in the building industry, one million are self-employed, working for themselves within their own construction business. They are usually one-man bands. They might have a mate or a few other lads on board. They are the glue and the gel that plug the gaps of the building trade, doing the little jobs, and earning themselves a weekly wage. They maybe operate with just a couple of regular clients to keep them going (e.g., landlords or small contractors) on a weekly basis.

A lot of the work these guys come by stems purely from word-of-mouth recommendations. I saw a bloke at the football just the other day. I've done a bit of professional work for him, and I asked him how he was getting on.

He's a joiner, and he'd just finished the refurbishment of a luxury house for a private client. The job had been a bit of a nightmare as the money-minded owner had cherry-picked from his quote (one that I had prepared for him). Rather than just showing the overall price, he'd gone open book and shown the client the full breakdown, demonstrating how the total price had been derived.

Result? Whenever the client saw an item on the schedule of work that he thought he could get a bit cheaper elsewhere, he took that piece of work off the to-do list. It meant that the joiner was left, basically, with all of the crap. The stuff that was probably underpriced or undervalued.

(Okay, I'd helped him price it, so you could say that it was all my fault, but we'll come to that shortly!)

Anyway, it was a thankless, unprofitable, and unhappy task for the joiner for many months. Then, having seen the job to fruition, the client asked the joiner to fit out three cafes that he owned. It was a chance to make back some of the money that he thought he'd lost. The joiner didn't complain about being short-changed on the initial work, he did a good job under the circumstances, and he then got called back and offered the chance to realise some of the profit he'd lost on the luxury house items that had been removed from the original scope of works. So, it all worked out in the end.

At the same time, this young builder had taken his son to junior football practice where – while watching on the sidelines – he got chatting with one of the other dads. The other guy was a surveyor like me, so they started talking construction business and at some point, my name came up. Both of them knew me; now they were networking, with a mutual acquaintance in common.

So, when I bump into this guy at the match and ask him how things are going, he tells me that he's now getting work off this old colleague of mine. That's how it is for these little guys.

Small jobs, word of mouth, bits of work leading to other work. In fact, small contractors are forever putting work each other's way. If they have the phone number of a couple of brickies, sparks, chippies, plumbers, plasterers, painters, scaffolders, roofers, glazers, carpet fitters, groundworkers, and tilers – and if they can get all of them onsite at the right time and in the right sequence – then who needs a large or even a medium-sized outfit, with all of the overheads that go with it? An orchestra of little guys can work just as well.

A bloke once said to me, 'It's hard to make half a million quid on your own. But get ten blokes, or ten small contractors, all pulling in the same direction, and giving each other work, and you can make that half a mill.' Just remember it belongs to all of you. Be sure to share it out.

There's a whole ecosystem taking place down there at the smaller end of the construction industry. Looking after the pennies so that the pounds take care of themselves.

These are the little guys.

THE MEDIUM-SIZED GUYS

Let's say the little guys in construction employ between one and 20 people. By comparison, medium-sized firms employ between 20 and 100. Again, they are everywhere in the industry.

They'll have some management people, some admin and accounts personnel, and possibly a receptionist or two. They'll have a certain level of retained staff in various trades. The rest of the work they'll subcontract out in order to complete a particular contract.

They're a business, and they will need continuity of projects in order to satisfy their overheads. If they have too many people on retained wages or on salary, and the work suddenly dries up for a month or two, then they've got a major headache on their hands. You need cashflow. A reserve in the bank. You have regular outgoings that need to be met. Your staff need to be paid.

Medium-sized firms might have a couple of ongoing contracts with the local council or a large employer or manufacturer in the area, and these ongoing arrangements will be their primary source of income. They might have a finger in several pies, of course, maybe doing the odd bit of development of their own (having spotted gaps in the market) or some private client work.

While the rewards of moving up the food chain may prove enticing, the risks attached grow in equal proportion. I recently visited a construction site for a pre-commencement meeting and was introduced to a potential new subcontractor. I told him a little about our company ("We're great. Come on board!"), and he told me a little about his outfit. He said they employ a total of ten men. Three of them, including himself, are company directors. Now, you might think that with three chiefs and only seven Indians, that headaches might be few and far between. After all, with so much supervision in place, what could possibly go wrong?

Not so, as this bloke explained to me. The three directors are all still working on the tools as well as managing the company. These three bosses constantly fret over keeping their labour force occupied and earning their keep. The company is therefore happy at the size they are at.

A friend of mine, by contrast, has recently branched out on his own as a subcontractor rather than a sole trader. I tried to tap him up to put him on the books of my current company, but he seems to be just fine with the clients he currently has. He employs two gangs of four men. He did have three gangs, but it proved too difficult to control. Too many people were depending on him for their weekly wage, and the logistics of keeping them busy, ordering the materials they needed and organising the deliveries of the same to feed their endeavours was all a bit too much for him. So, he scaled it back down to the two gangs. That's as much as he thinks he can handle.

Still, there are a plethora of companies that occupy the middle ground when it comes to construction. Here's what they do.

If we say that a small company might turn over anywhere between £100k (for a sole trader) and a couple of million annually (for ten or so staff), then a medium-sized outfit might operate in the realm of two to ten million. If they have 50 employees at the higher end of those figures, then that means each operative is generating £200K per year. This will cover their wages of, say, £50K, plus all of the backroom support, the van, phone, and petrol they use, the computers and phone lines and internet in the office, and all of the payroll, accounts, and admin that they need. Plus, it will leave the company a margin known as hoped-for profit. Quite important that last bit.

The company might be family-owned or run by a collective of like-minded individuals. You would expect a degree of experience from those at the top of the tree. They'll have trusted foremen and women, a reliable workforce (if they're lucky), and they'll tender for opportunities in the construction market with preferably some repeat clients and repeat business.

They might operate in any or all of the key sectors in the industry, including health, leisure, education, residential, commercial, industrial, retail, or transport. They might do internal fit-outs, new-build work, extensions, refurbishments, or specialist stuff such as structural or restoration work. Their projects could range in value from £50k to £5m. A couple of the latter each year will see them comfortably reach their desired turnover. A few of those smaller jobs might tide them over while waiting for the big jobs to begin. Good little filler jobs. Plugging the gaps in the calendar.

These medium-sized construction companies will usually manage a lot of their tasks in-house. This can provide reassurance for a client as – should they ever have any complaints or hit any problems – one phone call or email to the boss should (in theory) get things moving and take any log jams out of the equation.

The company might also offer a variety of additional services such as design or planning, or dealing with utility providers, getting those all-important services to site, and all of the associated gubbins that go into making a construction project whole.

They'll have a head office, with possibly a satellite unit somewhere strategic to their operations. They'll have four or five vans bearing their company logo. It's all good advertising, after all.

In times of recession, they may feel the need to scale things back, cut their cloth accordingly, and become one of the smaller guys for a while. Or, fingers crossed, things are going well and they might decide to step up a gear into the big league. It's a brave move, and not for the faint-hearted...

Those are the middle guys.

THE BIG GUYS

The biggest construction company in the UK is Balfour Beatty. Their turnover in 2021/2022 was touching £9bn, up 8% on the previous year. The second highest-ranking building firm in Britain is Kier, who have (apparently) almost gone bust in recent times.

I've previously worked as a surveyor on a project so large that it was run as a joint venture between both of those two major outfits. The project manager was a 'Sir', so prestigious was the work. The honoured gentleman would give pep talks at regular intervals to all of us 'black hats' as we – the managing personnel – were referred to. I'm sure we get called a lot worse onsite!

I remember one particular talk that he gave when he told us (in no uncertain terms) that we *were* going to finish on time. We just were. No ifs or buts. We were going to do so, with no room for argument whatsoever. We would all just have to produce.

That's one way to get things done. Leave no room for failure. Failure is not an option.

You see, these big companies get handed work of national importance. Build a new railroad, motorway, or a national football stadium and the eyes of the country are upon you. The media take an acute interest in what you're doing, and it's going to be headline news if you screw it up.

There's good money to be made here at the top end, but it's often taxpayers' money, particularly if you're talking infrastructure. Taxes affect you and me, so we're all stakeholders in this. We're all watching. We're all concerned.

The big companies employ people in their thousands. They operate regionally or nationally, and they bring huge buying power to the table. When they pick up the phone, suppliers generally jump to attention and pull out all the stops to secure an order.

Let's say you have a construction project worth £100 million. Typically, 20% of the delivery costs will be in the management and the supporting apparatus (e.g., the skips, scaffold, and welfare facilities, the site compound, site cabins and the like). The breakdown might then be a 50/50 split between labour (i.e., the operatives' wages) and a similar sum for the materials that the workforce are going to use to construct the actual building.

You could be picking up that phone and buying £40 million worth of gear. Big numbers. And any supplier worth their salt should want a piece of that.

Large construction firms tender for work, alongside their competitors, and are (normally) suitably resourced to manage a project from conception to completion using their own in-house staff.

On their books are planners, designers, estimators, project managers, and site managers, as well as setting out engineers, quantity surveyors, and legal experts. They'll generally work on projects ranging from one million in value up to £100m, and sometimes even more than that.

They'll have lots of projects on the go at any one time, with their own buying department placing those much sought after orders for materials. Money gets spent hand over fist to keep the people onsite stocked with the goods needed to carry out all that physical work.

All of their activities will ultimately be overseen by a director of operations, a managing director, and a commercial director. They'll probably also have their own finance and accounts team.

Some of their work might be years in the planning, and they'll need a whole avalanche of information just to be able to quote for the work in the first place. Their team of estimators will need to pore over every detail to make sure they don't miss anything that should be included in the price that they will ultimately submit to the client.

As the size of the job goes up, the expected profit margin should, in theory, come down. If you're doing a job for a thousand pounds, you probably want to make a few hundred quid on it. £500 profit on top of £500 outlay is double-bubble; that's a 100% mark-up. You can't do that on a £100 million job.

You might drop down to a 10 or 15% margin on a job of that size. That's still nice money if you can get it. But you have to get the price right at the very start; otherwise, if you go a degree or two off-course along the way, the project can end up costing *you* millions instead of making you that hoped-for profit.

For example, the new Wembley stadium was a done deal at £750 million, but then it cost the contractor over a billion to build it. Oops! Turnover is vanity, profit is sanity.

In theory, bringing in a big firm for a big job ticks a lot of boxes. The staff should share some sort of company ethos, and they are all operating on the same side, so there shouldn't be too much infighting within the project team. These are, or at least they should be, can-do companies.

And then sometimes – even with these big companies – things can and do go wrong.

I'm thinking Carillion, obviously. If it can happen to them, then it can happen to anyone. But I've also known a couple of medium-sized companies go bust in the last couple of years.

In the North-West (my home base and therefore an area of particular concern for me), we recently lost both Bardsleys and CPUK. I personally wasn't really surprised at the latter's demise. I'd dealt with them on a couple of occasions and, each time, I was simply left aghast.

You couldn't get a decision out of anyone there. Everything that you presented to them, including invoices for work carried out, would go through about six or seven levels of management approval before you got paid. If it takes six or seven people to sign off on a piece of paper, then who the hell is out there running the actual jobs, which is the most important part of the business?

That said, I'm often in awe of some of the great British construction companies. Many of them are populated by hugely talented people who do amazing, jaw-dropping work, and I'm often at their table, hearing them think aloud as they set about transforming Britain's landscape.

Big construction can be a huge force for good. It employs and it delivers. And, as long as it doesn't drown in middle management, it can make a reasonable return on those billions of pounds that pass through its cash registers.

That's big business. That's big construction.

6. DIFFERENT CONSTRUCTION ROLES: PART ONE

HANDS OR HEAD

Say you're drawn to a career in the construction industry. Good for you; we need people. There are so many opportunities there, and the variety of those opportunities is amazing.

The first thing you need to decide is whether you want to work with your hands or with your head. I'll qualify that a little. Craft workers are extremely intelligent, and they use their heads as well as their hands, but as I wish to talk in general terms here, please forgive me for adopting a general rule of thumb.

It seems to me that anyone about to join the construction industry is immediately faced with a fork in the road. Do they want to work with their hands – on the tools as it were, doing the physical grafting – or do they want to work on the management and/or administrative side? That's the basic choice facing each and every one of our new candidates right there.

We'll talk about the ways to go about achieving both in due course, but I just wanted to talk a little about the basic differences between these two potential avenues.

Some people see a respectable, fulfilling, and sustainable career in the skilled work areas of construction. They train to be carpenters, electricians, plumbers, plasterers, brickies, and the like.

Maybe every day is different, working on a variety of construction projects, and every day is demanding in its own way. These craftspeople work up a sweat all day long, and at the end of their shift, they are able to see the results of their endeavours. They can then go home, put their feet up, have their dinner (or tea), watch a bit of telly, and spend time with their families. Then they go to sleep and wake up and do it all over again the next day.

They are craftsmen and women. They work with their hands. This is an option for you as you consider your career in construction.

These skilled men and women are also supported by a host of journeymen: labourers, hod carriers, odd-jobbers, and those that don't have the necessary paperwork or apprenticeships to class themselves as 'skilled' (but nevertheless support and do the fetching and carrying for those who ply their hard-earned trade).

This is the physical stuff. This is what most people think of as actual 'construction'.

Then there's another side. Maybe one that a lot of people don't realise exists, or certainly don't realise its importance to the building industry. This is the management, supervisory, and administrative side.

A colleague once said to me, the guys onsite don't have 'our mentality'. What he meant was the onsite team didn't understand the behind-the-scenes pressure to ensure a successful and profitable outcome to a project. As long as these workers ended up with a decent day's pay for a day's work, their world was in equilibrium.

I had a job once where we employed a team of subcontractors to do some roofing work on a factory unit on a large industrial estate. The company I worked for was small, probably turning over about £2m a year, max. They had no contracts manager on board (i.e., no one overseeing the work being done on our behalf).

The gang of subbies, therefore, had no supervision. There was no one to ensure that they jumped out of the van at eight in the morning, put in a decent shift, did the job right, and then clocked off at four in the afternoon.

What did this unsupervised lot do?

They turned up to site about 9.00. And just sat in the van. Got out of the van around 10.00 and got on the roof. Messed around until 11.45 or so, then got back in the van and had lunch until about 1.15. Then they did another hour and a half of so-called graft before heading home about three.

It was the best job they'd ever had.

The site manager of the entire industrial estate rocked up one day. He could smell cannabis in the air, and saw no work being done.

There had been almost nil progression since his previous visit to the project a month earlier.

Our company got kicked off the job. Our client, who basically had the entire maintenance contract for the whole industrial estate, was similarly banished.

My boss decided to let the subbie gang go. It was a few weeks before Christmas.

Their main man complained. 'I've got presents to buy for the kids. How am I going to tell my missus I've been let go just before Christmas?'

Well, you reap what you sow. I don't wish to be heartless, but they should have had a go when they had the chance.

And the failure of that project buggered up so many professional relationships and the loss of several very profitable contracts. They don't have our mentality, as my colleague said.

This story illustrates why we have competitive tenders. That's why people get given a fixed price to deliver a certain scope of works. An open chequebook generally does no one any favours.

As for management, many people who have worked 'on the tools' end up overseeing the work that they formerly did themselves. I've heard it a hundred times – 'I used to work on the tools'. It makes sense.

Certain (perhaps all) manual work comes at an eventual cost to one's physical condition. Laying bricks is hard and carrying lengths of steel and timber is hard. Plastering walls, and probably every physical task you can think of becomes that much tougher as you get older.

Plumbers and electricians have to dig into floor joists and walls to lay their pipes and cables. They're up and down, 100 times a day. That's no good for the knees.

If you qualified in your trade in your early twenties, then a quarter of a century later – by the time you're in your mid-forties – you're probably feeling the pain in your joints.

Contract management, or maybe becoming a working supervisor, is therefore a logical option. You can use your expertise, and although there's a lot more paperwork, and a bit more responsibility, at least you live to fight another day.

Some of us go down the pen-pusher route right from the very start, of course. Which often prompts those onsite operatives to declare that you've never done a proper day's work in your life! (I started out as a labourer, so I've got my retort to that last accusation well prepared.)

There are many opportunities for professional and administrative staff in construction.

We need buyers, planners, designers, estimators, and clerical staff behind the scenes. We need management personnel to oversee quality, health and safety, and to protect and maximise the project finances once these schemes go live onsite.

Thinking of the construction industry as a potential career?

Ask yourself first, are you better working with your hands or with your head?

Whatever answer you come up with, the industry wants and needs you all.

TRAINING AND QUALIFICATIONS

So, you've decided the construction industry is for you. You've met that fork in the road and chosen between head and hands. So, how do you get from there to where you really want to be? Well, training and qualifications.

In the olden days, that would have been a straightforward apprenticeship. Basically, you would work under the tutelage of an experienced old hand, with some monitoring and guidance, periodic examination, and you would spend about four years learning your chosen craft. You would still learn and improve thereafter, of course, but you'd be qualified, you'd be earning a proper wage, and you'd be let loose to ply your trade in the construction world on behalf of your clients.

But apprenticeships, in the real trades, are not as common as they once were. Nowadays, it all seems to point in the direction of college courses. NVQs, BTECs, or something called the T-Level, which came online in 2020. They all offer practical qualifications.

Through a combination of classroom theory and practical experience, you can earn an industry and employer-recognised qualification that will give you the right to call yourself a qualified tradesperson.

I know a bloke who is a signwriter. He left school at age 15 and got a job as an apprentice. First task, make a sign for the men's loo. First attempt, woefully dismissed.

'If you can't get the height, the spacing, the detailing just right, you better fuck off now,' his mentor told him in no uncertain terms.

Would anyone dare say that now? Actually, in the often-robust world of construction, they probably still would!

The apprentice signwriter's working day would always begin with him making a cup of tea for both his mentor and himself. One day, in a moment of mischief, he took his boss's favourite mug, screwed it firmly into the workbench, and then filled it to the top with the older gent's daily brew.

'Cuppa there,' he said casually when his boss strolled into work.

He then watched as the man struggled to lift said cup off the table on which it sat. The penny eventually dropped, and you can imagine the cursing that followed!

Anyway, after five years of learning, doing, and doing again, the apprentice became a craftsman. He could actually remember the day that he 'graduated'. They used to have a lorry turn up once a week with all of the materials that they needed to make their signs. It was the job of all of the young lads to unload the lorry, doing the heavy lifting of the metal sheets, pieces of timber, and tins of paint.

One day, the lorry arrived, and the foreman of the factory asked the apprentice to join them in unloading the goods. He was about to do so – a task he'd performed every week since joining the firm – when his mentor told the foreman that he couldn't have him.

Asked why, he said it was because the lad was no longer an apprentice. He was a signwriter, and he no longer needed to perform those menial tasks. The newly inducted craftsman could now be left and trusted to stand on his own two feet and provide a professional service in the name of his trade.

There was no passing out parade or cap and gown ceremony, complete with expensive choreographed photo opportunity to acknowledge his achievement, just the endorsement of the master craftsman. That was all it took.

In today's world, though, you need that piece of paper to back things up – City & Guilds, NVQs, BTECs, or capital Ts. For electricians and plumbers, their work needs to be signed off by a fully accredited engineer. It's not enough to be good; you need the authority to sign it off, too.

College courses currently offered in the construction trades typically involve almost 2,000 hours of practical and theoretical work. The other 8,000 hours to become an expert (vis a vis the 10,000 hours espoused by K. Anders Ericsson) will, therefore, be earned on the job.

As for professional qualifications, a HND or a university degree comes to mind.

When I stumbled upon my chosen career, I was in my mid-twenties. A little research told me that the quickest way to qualification would be a foundation year in college followed by a two-year full-time university course. Three years out of my life.

I had no dependents. I could do it. But, three years! I'd be 28 by the time I finished.

In a moment of clarity, I told myself that I hoped to reach 28 at some point anyway, so did I want to get there with those qualifications behind me or without? I sucked it up, had a great time doing it, and had the letters after my name to propel me into a bright future of 40 prospective years of decent technical and professional work. Not that anyone has ever asked to see my university certificate, by the way!

PROFESSIONAL BODIES

You might be good at your job. You might have a good little business in the building industry. But how do people know that? How do they find you? What accreditation and, therefore, credentials do you have?

There are several bodies that can provide the desired support and affiliation. The Royal Institute of Chartered Surveyors, or RICS, is one. They deserve their own separate section. It's coming next, especially given recent events at that organisation. (Spoiler alert – juicy gossip coming!)

Then there's the Chartered Institute of Builders. Their website proclaims them as the world's largest and most influential body for construction management and leadership. They even offer a *Master* version of their membership. When you're handing out your business cards, you could have the letters CIOB or even MCIOB after them. If nothing else, I think it shows prospective clients that you take your day job seriously. If you've gone that extra mile to join, and jumped through hoops in order to get there, then I guess that's a statement of your professionalism and commitment.

I once worked in private practice surveying for a boss who had about ten certificates on the wall behind his desk. He was a manager of this, an expert at that, an affiliate of whatever, and an associate of everything else.

He never said hello in the morning. I'd walk into work, sit down directly opposite him, and he would not say a thing. I used to think, surely rule number one in the management handbook is *greet your staff*!

Some councils and organisations will only work with lettered people. After all, these corporations and authorities are accountable to their stakeholders. If things go wrong, at least they can show that they tried to find trusted and qualified people to do the work.

Membership of a professional body can make you stand out from the crowd. Not everyone joins. Not everyone has to. Yet, it does give an air of respectability, and there's nothing wrong with that.

THE R.I.C.S.

The Royal Institute of Chartered Surveyors. Sounds good, doesn't it? These are the standard bearers, the crème de la crème of our profession. After all, we bear a royal charter. What could possibly go wrong?

Well, a recent scandal has rocked this noteworthy institution. To the point where its members might wonder what they have actually bought into. Will they be guilty by association? Would it even be better not to be a member at all?

The organisation can trace its roots back to 1792 when a group of like-minded gentlemen sought to pool their expertise and advance the reputation of their profession as surveyors in the construction industry. It all sounds very Phileas Fogg, but almost a century later – in 1881 – they were awarded a Royal Charter. Later, in 1947, they officially announced themselves as the Royal Institute of Chartered Surveyors. All very grand.

RICS offers technical-level membership as well as that of Full Member, Associate, and Fellow.

Many employers demand accreditation as many clients and organisations, such as insurance companies, will not take the word of any surveyor who doesn't carry the RICS letters after their name. With it, you come fully bonded. Without, well, you could be anyone.

An august operation, then, no doubt.

One that was recently let down in a big way by its leadership.

In 2018, a financial report was carried out by the accounting firm BDO that found serious governance issues. RICS had well over 100,000 members globally with an annual income of £80 million, and just four people (including the president and chairman) had ownership of those funds. And those four people answered to no one. They could do with those funds whatever they wished. That's what the report said.

Also on the board of the RICS were four non-executive members who, upon reading the BDO report, recommended that it should be shared with the rank-and-file members. 'Put it out there,' they said.

Guess what? They were all dismissed.

The report had highlighted how the organisation was at risk of 'unidentified fraud, misappropriation of funds, and misreporting of financial performance' but, clearly, it was best not to tell the membership about that.

A subcommittee of previous board members, known as the GC19, also raised their alarm at the report's conclusion and the treatment of the (dismissed) non-execs.

Cue more entrenchment from the gang of four. According to BDO, these gilded individuals didn't like being challenged and they were also quick to take offence when criticised.

Now, with the non-execs free to air their grievances to the press, and the GC19 members doing likewise, attention turned towards the organisation itself that carried that royal seal of approval.

QC Alison Levitt was brought in to carry out a review of the crisis. She concluded that the RICS should make a public apology to the ousted directors who had raised the alarm when the initial report was essentially buried, and also apologise to the GC19.

She said that both of these parties had behaved in an exemplary fashion and that they were right and brave to have raised the issues in the face of quite a lot of hostility from the RICS board.

Huge shame on the latter.

The construction magazine *Building* reached out for opinion and found that the core membership of the RICS felt 'disengaged and neglected', and overall satisfaction with the organisation, based on 9,000 responses, was at just 43 percent.

At present, Lord Michael Bichard has been appointed to carry out an official review that will examine the whole purpose of the RICS. Maybe it will lose its royal charter.

The people responsible for the scandal have all resigned, praising each other as they left the building to pursue other opportunities, having had a wonderful time, and being immensely proud and deeply privileged to have been able to serve this wonderful institution, etc, etc.

So, what went wrong?

I once dealt with a regional director of the RICS in his personal capacity as head of his own construction company. According to the RICS, he had held the position for a decade or more and was known for his 'integrity and fairness'.

The fucker then lied to me!

I was chasing down a debt. We'd done some work for them, and were owed money. I'd only recently joined the firm I was with, and I asked if there were any outstanding accounts that I could try to tie up on their behalf.

Pointed in the direction of a job we'd done for this bloke, I read through the job file, appraised myself of all the issues, put my 'fair hat' on and walked through the bones of contention.

One of the heads of claim should go to us, I thought. One of them should go to them. Meet in the middle on the next one, that sort of thing, until I had a settlement figure in mind. It was time to give this fella a call.

I told him who I was, who I represented, and explained that I was trying to reconcile our account which – even with my 'fair hat on' – looked like they owed us a few quid. Not loads. Maybe ten grand.

I explained my credentials. An MA, I said, with 20 years of surveying experience. I'd also written a few books, including some on construction. I know what I'm talking about was the basic message. Don't mess with me.

'And I'm the president of the so-and-so region of the RICS,' he said haughtily. 'Have been for 15 years.' Something like that.

I asked for our money. Even taking into account all of his previous rebuttals of our claim, I believed we were owed this amount of money, and mapped it out.

He acquiesced. He said there were a few residual issues on the job and that if we would correct them, he would issue a payment for my fair settlement sum.

Happy days.

We went back to site, did the repairs, and I then called him up to ask for our money.

You can guess what happened.

He moved the goalposts, saying that because of blah, blah, blah, he wasn't going to pay because of bullshit, bullshit, bullshit.

'Listen,' I said, 'that is not what you and I agreed to. You said that if we did x, y and z, you would then release those funds.'

Now, he wanted us to do some additional work, and only then would we get paid.

He'd lied to me. He was taking advantage. His word was certainly not his bond, and the matter ended up in arbitration. Naturally, this master gamekeeper turned on the charm and knew what to say (he'd sat in that judge's chair himself on many occasions), so we were on a hiding to nothing. My boss lost his (in my view) legitimate claim.

That was a president of one branch of the RICS.

And as for the organisation itself? Well, Lord Bichard's report will tell us more, later in the book.

7. DIFFERENT CONSTRUCTION ROLES: PART TWO

THE ARCHITECT

Architects are the mad professors of the construction industry. They are a delightful combination of both the creative and the technical, and their role is often to come up with the weird and the wonderful. They really do bring something different to the table when you are generally dealing on a day-to-day basis with grey and unimaginative business people or thick-armed (and sometimes thick-headed) building folk. For me, architects are pretty much geniuses, one and all. Their technical knowledge never fails to impress me, their creative input is always valued, and their ideas are often surprising, unexpected, and inspired.

Now, that's not to say that these people are infallible. They are only human, after all.

Some of them do get carried away from time to time. The garden shed you've asked them to design might end up looking like a miniature version of the Guggenheim Museum. Still, they see things that you or I might miss. They think outside the box. They can give you the best use of whatever space you are trying to create or remodel. The only downside is that they sometimes appear ignorant of the economics involved. The word 'budget' doesn't seem to appear in their vocabulary. In fact, it is almost like Kryptonite to them.

Many architects are self-employed or work within small private practices. In fact, three-quarters of all architectural practices have less than ten employees.

These creative professionals are initially approached by a client and receive a 'brief' from them. This might be a local authority with a plan for a new school or sports hall. It could also be a wealthy individual who wants to build a bespoke luxury home.

For example, Mr Super-Rich Premier League footballer wants a new six-bedroom house with four en-suite bathrooms, a detached triple garage, an indoor gym, and an outside swimming pool.

He's bought a plot of land on which to site the property and gives the architect an idea of his budget. Say he has £2 million to spend. Now, it's over to the architect to grill the client further as they dig into the details. Style of house, do they want lots of natural light, and how about a man cave for the footballer and his friends/hangers-on?

Suitably briefed, the architect can get to work.

They'll probably use a rough cost/per m^2 to try and stick roughly to that £2 million budget as they design the overall footprint of the plot. The rest is then straight out of the architect's head.

Let's take a project like the previously-referenced Shard in London. It's only the tallest building in the United Kingdom, and the client was an entrepreneur called Irvine Sellar, whose unique and personal vision provided the impetus to drive the project from start to finish. Eventually completed in 2012, it all began with an informal meeting between Sellar and the Italian architect Renzo Piano back in the year 2000.

Both being international kinds of guys with busy diaries, they had their first design meeting at a restaurant in Berlin.

The client explained the reason for the invitation – he wanted to build a 400-metre-high tower in London… at which point the architect groaned. He said he hated those non-descript tower blocks; constructions like Canary Wharf and their ilk. They looked like nothing but symmetrical vertical boxes.

It was once said of the ill-fated towers of New York's World Trade Center that they looked like the boxes that the Empire State and Chrysler buildings came in!

But a brief is a brief. A client is a client. So, grabbing a napkin, Renzo took out his (probably very expensive) pen and started to scribble a design. The result resembled a church spire that he'd once seen in a painting that reminded him of old England. Or even the mast of a ship.

The site for the tower was right next to the River Thames, so the ship theme was appropriate. And the spire theme would make the building different to everything else that was being built. After all, a tapering design would mean less space on each floor as the building ascended to the heavens. Most schemes are only ever approved where profits and rental returns are maximised, so the client was risking his rewards by reducing the overall available rental area.

When the initial designs were properly drawn up (i.e., not on the back of a napkin) and presented to the planners in London, one detractor claimed the building looked like a shard of glass jutting into the sky. Although meant as a negative comment, the name was gratefully – and cheekily – adopted by the project team, and The Shard project began to take shape.

I've been in the building several times. With a great location overlooking the Tower of London and all the major landmarks, it offers amazing views (as you would expect) over the city. In fact, what's not to love about it?

There's a story that when the building was finally completed, and prospective tenants were being invited in and shown around, Irvine Sellar would always step away from the visiting party and go and stand by a south-facing window as if deep in thought.

Out of curiosity – as he was by all accounts an inspirational and interesting guy – people would be drawn to his side to find out what he was looking at.

Then, to clinch the deal, he would say to them, 'And to think, on a clear day, you can see all the way to Paris.' It seemed to do the trick, as all of the building's 72 floors are now fully occupied.

If only they had known that what they (and Irvine) were looking at was the Crystal Palace communications tower!

This is not to say that all schemes are so grand and so big. Architects can put together an extension to your home, offering a design that makes best use of your space. Give them some pointers, then see what they can come up with. Like I've said, they just *see* things.

I used to play football with a bunch of architects, and it wasn't unusual for us to talk a bit of shop at the pub afterwards.

A historic building in a city centre location had recently become home to a housing association, where I had just been to a client meeting. Huge ornate metal gates sat at the junction with the pavement, followed by a large courtyard and then glass revolving doors leading into the offices themselves.

As there were also offices above the empty courtyard, the area was dark and foreboding and didn't make a good first impression.

So, what did the architect do?

Well, they tried to think of a dark and dingy place that could also be inviting. Even magical. They turned the courtyard into a grotto, complete with a winding path that led up to the revolving doors. Plants and trees, as well as fairy lighting on either side of the winding path, completed that enchanting entrance.

It put a smile on the face of practically everyone who passed through, and turned that previously-foreboding space on its head. It created happiness from misery and light from the dark.

When I mentioned how surprised I'd been at the transformation, my architect friend said, with great satisfaction, 'I did that.' I'd complimented him on his own work, when I hadn't known that he was responsible, and the subsequent smile on his face was an absolute picture.

That's architects for you. The very best of construction.

But here's another side of the story. Just to balance the books, as it were.

I once visited a project where four new houses were being built on the site of a former grain store out in the countryside at the edge of a wealthy commuter town on the outskirts of Manchester.

We were going to be doing the roofing, so I turned up to view the job and see when they might be ready to take delivery of the slates and the labour to top these houses out. I wanted to check that the walls were all up and the roof timbers were in place, that sort of thing.

I went into the site manager's office and introduced myself as the rep for the roofing subcontractor. I then asked him how it was all going. 'Nice houses, by the way,' I added. I'd previously seen the drawings. Now I could see them in the flesh, as it were. Four large detached properties. Desirable homes. Bound to sell easily.

'Yes,' he replied. 'It's just a pity they're the wrong way around!'

What he meant was that the *front* entrance of each house was facing out into the distance, to the countryside beyond.

The gardens of the four properties were all conjoined at the back, like a square divided into four equal pieces. Obviously, there would be a dividing fence, but imagine if you were having a barbecue, or sitting out back with a glass of wine and some music playing on a balmy evening – you'd be overlooked by all of your neighbours.

Assuming the bedrooms are at the back of the house, as they often are, then when you throw the curtains open of a morning, dressed only in your boxer shorts or dressing gown, you might find yourself staring at your neighbour opposite doing the exact same thing at the same time. Hardly private, is it?

If those properties had been spun 180 degrees, the scenario would have been completely different, and wholly more desirable. Park out front and wave to your neighbour as you come home from work, then go through the front door into your own personal kingdom. Lots more privacy, and gardens out back overlooking the countryside beyond.

I'm not quite sure what the architect was thinking here, but they definitely dropped the ball on this one. And once buildings are up, they are a lot more difficult to turn around!

So, an opportunity missed on that project. The result was a less desirable home for the people that eventually bought them. You can't win 'em all, I guess, but I do still love architects.

In a not dissimilar vein, I read a story in *The Quantity Surveyor's Bible* about a new-build apartment block that was freshly springing up from the ground. The setting out of the foundations was all laid out to exact coordinates as arranged by theodolites, satellites, and choreographed on the architect's drawings.

And yet they were wrong. Only by a few degrees, but those precious few metres took the development close to a much-used railway line. The reverberations were going to rock those misplaced foundations to the core.

The project manager, who had clear sight of the correct line that the brickies and groundworkers should have been building to, instead chose to believe the numbers on the drawing rather than his own eyes and his own common sense, not to mention his construction experience. And then, he chose to blame the architect when – in reality – it was his mess as well. Asked what he was going to do about it, he responded, 'We'll just move the building.' All 100 tonnes of it. What, just grab a corner each, lift it up, and plonk it down in the correct spot? Far better to set it out correctly in the first place.

Check the positioning before you start to pour the concrete. Or pay the price.

THE BUILDING SURVEYOR

Ah, fungi!

That's what I always say to myself when I think about building surveyors.

It may be a bit unfair, but they seem to employ their undoubted talents in one of the least interesting aspects of the construction game.

For example, no building scheme – nothing that is ever being built – has a building surveyor as part of the initial project team. Architect? Yes. Project Manager? Yes. Quantity Surveyor? Yes. Building Surveyor? No.

So, what is it that they actually do?

Well, say you have a crack in a building. Say you have some sort of mould on the wall. Say you have a leak somewhere. The building surveyor is the expert that you call to find the source of the problem.

In terms of planning your project, they can give advice on building regulations, minimum space allowances, party wall agreements, and the like. They are good at all of that stuff. They can advise on maintenance, they can identify the cause of defects, they'll take photos. Do you a nice report.

Basically, these are the nannies of construction. And I can't for the life of me see what floats their boats.

Okay, there are worse jobs in the world. They don't have to clock in and they're not stood on a production line. Probably no two days are the same. They can still get accredited to the RICS, and they are professional and responsible people.

But construction? They don't construct anything. They advise on the *condition* of buildings.

When I was at university, there were 50 of us on our course. The first year was common to all of us: basic, industry-wide stuff. Then, at the start of the second year, we all had to select a major discipline. Five of my peers opted for Construction Management. Five of us went for quantity surveying. Forty of the 50 went for building surveying.

I mean, there's only so much damp to go around. Where do all of these people work?

Simply waiting for things to go wrong and then analysing and providing remedies seems to me to be a niche market. That said, I'm a big believer in the value of maintenance. If you look after something – whether that's your home, your car, or your body – then it's likely to give you a lot more bang for your buck and stand the test of time.

Like the guy who lived to be 100 and said, 'If I'd known I was going to live this long, I'd have looked after myself better!'

Anyway, if someone is thinking of buying a home, they might commission a surveyor's report. It could be a simple condition report, a homebuyer's report, or a building survey.

The building surveyor will generally carry out a non-intrusive assessment, looking for any signs of damp, structural movement, dry rot, evidential defects. All of value to a potential purchaser.

The building surveyor can advise on energy efficiency, the use of space, urgent repairs that are required, and also indicate problems that may arise.

In turn, the building surveyor might deal with planning applications, act as an expert witness in insurance claims, prepare dilapidation reports, find structural faults, and recommend solutions after that. They are generally good at keeping up-to-date with the ever-shifting statute of regulations. They might guide your project through a legal minefield. After all, they have nothing better to do. Did I mention that there's only so much damp to go around?

Mostly, however, they deal with improvements and remedial works to existing buildings. They survey what is already there and report back on their findings.

I remember watching a TV programme about a home extension that failed to meet the standard for building regulations. How my wife loves it when I have the remote control! Anyway, a rear window had been installed to the property, but it was not compliant with the requirements for fire safety. It was too small.

'But I can get through that,' the slightly-built female homeowner protested.

The appointed building surveyor patiently explained that the fire regs asked for windows to be large enough for a firefighter, wearing full protective life-saving equipment, to be able to climb through. The Building Regulations weren't based on her own slim proportions!

Building surveyors involve themselves with the maintenance, repair, refurbishment, and restoration of our current stock of residential, public, and commercial properties.

The work that they do can be of great importance, especially to tenants and homeowners who find themselves in dire need of their services, such as the residents of a new-build development, Agar Court, in Camden, North London. Despite being less than six years old at the time of writing in 2024, and the apartments in the luxury block costing between £750K and £900K, three of the new owners have since spent a collective £250K trying to get the developer to cough up the cash to repair their failing homes.

The list of defects includes: multiple leaks, cracked tiles on the floor due to the uneven floor level, bricks that can literally be pulled out of the walls and, to add to the misery, the smell of sewage in the en-suite bathrooms.

In all, more than 150 defects have been logged in the complaint, which is currently being passed between the developer, the insurer, the conveyancer, and the local council.

An independent report by building surveyors Structural Surveys found built-in structural defects and called the outer masonry (quite an important element of the build!) 'unfit for purpose'. Their report also said, in conclusion, that consideration should be given to the potential demolition and rebuilding of the entire development.

Okay, so it's not all about the damp then!

THE PROJECT MANAGER

Who is the project manager? What do they do? Why do you need one?

Well, obviously, this is the person in charge of a scheme once it goes 'live'. To be honest, they need to be involved from the very outset, putting everything in place *before* the job starts onsite. Ducks in a row.

I once worked with a project manager on a new-build £100m hotel. The things that he had to consider, and the obstacles he had to surmount, seemed overwhelming to me. I was somewhat in awe.

Site set-up involved road closures, cutting off pedestrian pathways in order to put up hoardings and facilitate deliveries, liaising with the planning department of the local council and the ultimate service providers, ensuring that adequate resources were in place at every stage of the development, and this was all before we had even put a spade in the ground.

Then, through their energy, expertise, and passion for the enterprise (true or feigned), project managers have to see the project through to fruition. While the client is king, the project manager, for a time, is the day-to-day boss.

Now, if like me, you're a fan of the aforementioned *Homes Under The Hammer*, you'll often hear the question asked of the new owners of a property about to be developed – 'Who is going to be PM?'

Sometimes, the purchasers will say 'I am' and, occasionally, these people are seasoned pros. But not always.

Some of them try to wing it. And oh, the tales of woe at the end!

Project management is like being the conductor of an orchestra, and sometimes things do fall into place. But, at other times, the sounds emanating are just awful. What you can hear at that point is the sound of inexperience. And pain. It's a kind of high-pitched wail! Certainly a lamentation.

There is a tendency for inexperienced PMs to come unstuck and ultimately stall their projects. Contractors, or subcontractors, or solo tradesfolk tend to run a mile when they sense or see chaos (or a standstill) onsite. What these hardworking people want to see is a well-oiled machine. They're here for the opportunity to make money, so they need a clearly defined brief for which they can provide a quote, and on which they hope to see a return.

If they are invited around to do a 'bit of this and a bit of that', and they can't understand what the person (supposedly) in charge is asking them to do, then you can bet your bottom dollar that these people will promise to get in touch, and the novice PM will never hear from them again.

Clear direction. Clear vision. Clear instruction. That's what a good PM gives.

I was once briefly seconded to a high-rise scheme in the North West of England. It was an 11-storey residential tower block being newly constructed. This was at the height of the economic and construction boom, circa 2007.

On my first day on the job, I introduced myself to the site manager and asked how it was going.

He told me that he'd been up a ladder the night before fixing a leak in one of the apartments. It sounded like a design flaw to me when he gave me the ins-and-outs of it.

'Why were you trying to fix it? It sounds like a design flaw,' I said. (I'm obviously an expert. A frustrated building surveyor, perhaps.)

'Because if I don't, he shouts at me,' came the reply.

'Who's he?' I asked.

'The project manager.'

Oops!

The bloke laughed.

When I asked him when they were due to complete the scheme, he said, 'We should have been finished six months ago.'

Moments later, the PM came marching into and straight through the site office.

Where's this? Where's that? Why hasn't that been done?

He was a real charmer.

No handshake or hello for yours truly. I was bemused, to say the least.

I then decided to take a walk around the site and into the high-rise apartment block itself.

Given that the job was way behind schedule, you might have thought that there would be a small army of tradesmen onsite pulling out all the stops to complete the job. Better late than never, that sort of thing. No chance. It was like a ghost town.

I eventually found a couple of blokes on one of the floors doing a bit of tiling in the bathrooms.

I said, 'Hi, I'm your new QS.'

'Have you met Mister Personality yet?' one of them asked.

'Yes,' I said. The lack of motivation and leadership on the project was palpable.

These were the good times. A boom time in construction. Lots of work around. Do you think anyone was rushing to work for this guy when there was money to be earned elsewhere without the hassle of trying to cope with the demands of this particular PM?

Well, the market suddenly crashed around 2008, and these apartments still weren't finished. And in that desperate time of economic uncertainty, the flats couldn't be sold, and the company that was building them – and who'd employed this disaster of a PM to complete them – ultimately went into administration.

The moral of the story is that if you decide to be your own project manager, make sure you know what you're doing. Lead the enterprise. Engage and enthuse your workforce. Give them good direction.

If in doubt, employ an expert. Remember, the good ones run the ship, while the bad ones have the power to sink it.

THE QUANTITY SURVEYOR

The quantity surveyor is often referred to as the accountant of the construction industry. I've been a QS for a quarter of a century. I know of what I speak.

I have to say 'accountant' doesn't do the role justice. Although we are primarily involved with (and responsible for) the finances of a scheme, we look at buildings whilst accountants look at books. It's 3D accounting, as far as I'm concerned, and your QS basically holds the purse strings for the entire project.

If a client (your Premiership footballer, for example) is having something built, then they have probably employed a contractor to carry out the work. If it's the £2 million house that's being constructed, no one is going to wait until the end of the job to get paid. After all, it might take a year or two to complete. The contractor wants paying monthly.

The builder puts in his invoice at the end of the month for the work that's been done, and if the client has a team around him (and they really should), they want to check that their builder's claim is correct.

That's where the client-side QS comes in. The contractor says that all of the brickwork is complete and wants paying, so the QS comes to site. If only three of the four sides are finished, or only one of the two storeys, the QS reduces the claim accordingly.

Now, you don't want to hold onto those purse strings too tightly. After all, the contractor needs funding. You can't bleed the job dry whilst – at the same time – you also can't overpay.

So, the QS checks the value of the completed work. He measures it and values it.

The contractor will probably have their own QS, and the two parties will haggle, agree or disagree. The money is slowly, and often reluctantly, shoved across the table from one party to the other.

Of course, the main reason you can't overpay is that the builder might go bust at some point, and hopefully not because *you* haven't paid him what he was due. Still, it happens from time to time.

If you've overpaid and the builder goes bust, then you need to get someone else in to finish the work. So, you'll end up paying twice if that money has already been handed over.

Tight (but not too tight) control of the purse strings. That's the QS role.

Now, there are many different types of QS. I've been the guy asking for the money, and also the guy paying the money. I've worked in private practice on behalf of clients, and also worked for small, medium, and large contractors asking for or dishing out the dough.

The quantity surveyor might be tasked with measuring quantities of materials for procurement purposes. You don't want to buy too much (and have too much waste) or too little (and have men stood around with no bricks to lay or paint to apply).

The QS might well estimate the overall cost of a project in order to tender for work or even advise a client on the initial feasibility of a scheme (i.e., whether they can actually afford it).

It's a varied role, and also an important one in terms of successful project delivery.

But then I'm biased. I'm a QS.

Also, I think it's important to mention that as the holder of the purse strings, that way temptation lies.

There's many a back-hander to be had in the world of construction. The QS is better placed than most to take advantage of any such pernicious opportunity.

This can take the form of the 'envelope full of cash handed over in a pub car park' in order to divulge details of rival tenders to interested contractors, while entire organisations can even be guilty of corrupt practices. In 2011, Sweett Group, the UK's only FTSE-listed quantity surveying practice was convicted under the Bribery Act for paying nearly £700K in incentives to procure a contract relating to a £63 million hotel project in Dubai.

A subcontractor I once worked with told me a story of how a former line manager of mine – a senior QS – invited him for a coffee and then uttered the immortal line, 'A bribe goes a long way, you know!'

He was seeking inducement to put him at the head of our tender list when it came to placing orders for the works we had on our books. Hardly subtle. And this was supposedly my boss!

Another time, I spent a good bit of time preparing a tender with a value of around £350K, only to lose it to another contractor by the miserly sum of three hundred quid. That's what is known in the industry as 'suspiciously close', especially when you consider the contractor who won the bid by the skin of his nose was the brother-in-law of the guy awarding the work on behalf of the client.

All I can say about it is this. If you are a QS, try not to be tempted. After all, your livelihood depends on people handing you *their* money. Very few people want to bestow such a privilege on anyone accused or suspected of corruption.

And, to be honest, most of the quantity surveyors that I've met or worked with are trustworthy. They are professional people. In fact, we probably have a reputation for being too banal, stuffy, and fastidious… to put it another way, a bit like real accountants!

Anyway, it's often said in the construction industry that a good quantity surveyor costs you nothing. They pay their own way. They'll get you every penny that you're owed, and defend your corner financially on every project in which they're involved.

I, for one, am totally in agreement. But then I would say that, wouldn't I? I'm a QS.

8. DIFFERENT TYPES OF PROJECT

NEW-BUILD PROJECTS

The construction industry is worth nearly £200 billion to the UK economy each year. It represents about 7% of the country's GDP and employs about 10% of the total UK workforce. About one-quarter of all construction work is in the public sector and about three-quarters of it is in the private domain.

Of all of this construction activity, about 60% relates to new buildings. The remaining 40% is involved with the refurbishment and maintenance of our existing buildings.

Obviously, new-build means developing something completely new. Building from scratch. This is our ever-changing landscape that we're talking about here. A new housing estate appearing at the edge of an existing town or village, a new office block, or a brand-new retail park.

In the world of construction, there are essentially three main sectors we need to consider. These are:

1. Commercial, including offices, retail, and sports and leisure facilities.
2. Residential, meaning houses and apartments.
3. Infrastructure, being the links between it all and the services that support them.

Commercial is responsible for about 45% of all construction work. Residential is about 40%. Infrastructure is about 15%. Then, within each of these areas, the split is roughly 60/40, as stated above, in favour of new-build development versus working with buildings that we already have.

Let's take a look at the residential sector. There were over 200,000 new dwellings built in Britain in 2022. Some 40,000 are individual houses. The remainder are apartment blocks, which equates to

11,000 multi-occupancy buildings with an average of 15 flats per block.

Lots of people like new-build properties. Once these new homes come onto the market, they don't seem to hang around for too long before they are snapped up by eager occupants.

You can see the builders still painting the external fencing, putting in the road markings, removing the heavy machinery and decanting from site when there are already cars in the driveways, children out playing in the street, and washing hanging up on the line. There is such pent-up demand for places for people to live that no sooner are they built, they are occupied.

Property developers are even offering bonuses to anyone who can identify land for sale or an opportunity they can bring to fruition. There's often a finder's fee available if you can spot potential in a redundant or available site.

Farmers, for example, are sitting on a goldmine.

Land!

If you happen to own some of it, and can acquire the requisite planning permission to build on it, then there's no shortage of interested parties who would take that piece of land off your hands.

That old skating rink or factory that has laid dormant for many years? You better believe that someone has their eye on it. Before you know it, timber hoardings will go up around the perimeter and site cabins will appear. Machinery will soon arrive and reduce the original building to rubble. Then, the ground will be levelled, and the construction of something new will begin.

A new-build opportunity means that the developer is starting off with a blank canvas. They get to choose the overall design and the specification of all the materials that are going to be used.

As they are essentially starting from scratch, they'll need to break new ground. They'll need a team of advisers on board to deal with planning and building regulations. Architects to realise their vision. There will be service providers to consult with. How close is the nearest water, electricity, and gas supply? What about transport links?

Is the development in keeping with the surrounding area? Is it in a conservation area?

I used to live in an apartment in the largest collection of Grade II-listed Georgian terraces outside of London. There had been riots in the area not long before I moved in and it was a bit, how shall I say, edgy.

Being so close to the city centre, people slowly began to realise the attributes of the location. The streetscape was quite beautiful, and filmmakers flocked to the area for its charm and its period authenticity. It was also close to the hospitals and universities. A good catchment area.

Then, all of a sudden, our much-loved parade of shops was being demolished.

Eight-storey student blocks began to appear on the footprint of our previous local amenities.

One advertisement even had the cheek to proclaim, 'In the heart of the historic Georgian Quarter...' to which I wanted to add, 'You built this bleedin' monstrosity!'

These particular schemes are generally soulless enterprises, and they are often over-represented in the marketplace; therefore, they are an extremely risky investment.

Still, there are undoubtedly advantages and attractions to all new-build constructs. They are zero VAT-rated, for one.

They should come with all the latest technology, be more energy-efficient, and should be maintenance-free for the first few years at least. The kitchen, bathroom, flooring and decoration should all be sparkling so you can relax and enjoy it for a while before it starts to look dated or tired.

Some people just like 'new'. You can't blame them.

Responsible for over half of construction activity, putting money through the tills and putting wages on the table, I quite like new-build too. But as for the quality, well, we don't build them like we used to!

REFURBISHMENT WORKS

Buildings become derelict. Buildings get old. The everyday needs of the population change. Old homes, factories, offices, retail spaces may have passed their sell-by dates, but they could still have something to offer. Maybe a prime location. A solid foundation and facade. Some attractive period features.

Yes, the original construction could be knocked down with something else put up in its place, but it's often possible to rejuvenate those former premises and breathe new life into them.

A lot of people, especially non-construction professionals, are surprised by the costs that are involved in refurb work as opposed to building something new. The numbers can be quite similar. There's certainly not the savings or price differential that some people seem to expect.

Allow me to explain why.

There is zero VAT on new building works, whilst regular construction works relating to refurbs attract the standard 20%. That 20% figure, however, drops down to 5% if the about-to-be-renovated building has been vacant for the previous two years. This is easily achievable for any proposed scheme, as you can spend that time getting all of your plans, permissions, and competitive tendering done, and putting your winning project team together.

As for the construction work itself, working within the confines of an existing building presents its own unique set of challenges. They come with their own restrictions, for want of a better word.

You aren't smashing the present construct down to the ground and starting afresh. You can't get gung ho about the project. You need to work with, and preserve an element of, what's already there.

You might have to take it back to the bare brick, and replace electrics, plumbing, and heating. Then timbers, plasterboard, plaster and paint are required. All of the things you would bring to the table with a new-build development.

So why bother refurbishing a building when you could create something brand-new for what is often about the same cost?

Well, location is the first thing that springs to mind; for example, you might want to refurbish your own home. You don't fancy moving, but the place is looking tired and a few things are starting to bug you. You might want an open plan kitchen-diner, you want to add a study. You want to refurbish what's already there. That's small-scale refurbishment.

Then there is urban redevelopment and refurbishment on a grand scale.

Battersea Power Station? Refurbishment and redevelopment.

The Houses of Parliament? Presently being refurbished.

Why? Well, they are both listed buildings for a start. You can't bulldoze one of those without a bloody good reason. They would basically need to fall down around you before you could start again.

The costs of doing up the Houses of Parliament are more or less unfathomable. The complications are immeasurable. Estimates range from £10 billion just for essential repairs to £50 billion for full-scale improvements. Timescale? Ten to 50 years. In truth, the building is ready for the knacker's yard. But what can we – as voters and taxpayers – do about these eye-watering costs? Knock it down?

What would we build in its place? Certainly nothing so grand. We probably don't have the money and don't have the skills to build what our forebearers did. Sad but true. But the building itself is a *symbol* of our nation, of our democracy. An international icon. We can't ignore it, nor just let it rot.

To demolish it and rebuild something of similar stature in its place would probably take a quarter of a century at least, even assuming we have the ability to deliver something of similar quality.

Instead, your taxpayer pounds will (in all likelihood) be thrown at the almost impossible task of keeping the building upright. And, it must be said, our parliamentary home is a fabulous building.

So that's why we refurb. Things can be cheaper, yet it's often location and the desire to preserve what is already there. Sometimes, it's the best choice; sometimes, it's the only choice when listed status and restrictive covenants are in place. I'm all for the refurbishment of what we already have. It's greener, there's less manufacturing of new products, and less of the old stuff transported and dumped in landfills. More importantly, it's our heritage, and it's our future.

ALTERATIONS AND EXTENSIONS

Moving house is recognised as one of the most stressful things that many of us will ever have to do in life.

Now, if your current living arrangements are pretty dire and depressing, then you might actually be looking forward to relocating to new accommodation but, generally, most of us would choose to have a root canal rather than pick up sticks and move house. I know I certainly would.

There are many reasons why we might want to stay in the place where we currently reside. We like the house, like the area, like the neighbours. We could have family and/or friends nearby.

So, you don't really want to move. At the same time, the house that you live in is just not cutting it any more. You need an extra bedroom or an extra bathroom. Or how about a downstairs loo?

Still, there's no need to move, and plenty of reasons to stay put. There are no removal costs for one, or estate agents fees, or the fear of getting gazumped on any new property that might take your fancy. No, all you need to do is find a competent builder at a price that you can afford to make the alterations that you want. Then, you just have to put up with a bit of upheaval for a while. After all, you'll have the builders in.

A few months down the line (or longer, depending on the scale of the work), you'll have given your present (and hopefully happy) home a new lease of life.

But the cost of getting the actual work done is often the big sticking point.

As someone who regularly prices extensions and alteration work for small builders, as well as for private clients and homeowners, I don't think I've ever had a customer say, 'Ooh, that's cheap!'

Generally, the startled response is, 'How much! For a little bit of brickwork and plasterboard, moving a few electrical sockets... etc, etc!'

If you love where you live, and you just want to modify what you have, then you have to be prepared to put your hand in your pocket. From there, get yourself a few quotes, and do your due diligence on the paperwork that comes back to you. Don't just take the lowest offer. Cheapest doesn't mean it's the best of the bunch, or that it offers you the best quality. It never has done and probably never will.

Another thing to consider when thinking about altering or extending your existing home is the sheer amount of disruption involved. Can you afford a temporary rental? Maybe you have a second home in France or Spain that you can visit for the duration of the works. If not, you might be in for an extended stay with your parents or the mother-in-law.

Even that could be preferable to staying put while the work goes on around you. How do you keep your mouth shut while a bunch of strangers *do things* to your beloved home? Precious features can become mere obstacles to someone who simply wants to get in and get the job done in the shortest time possible.

The more you grow irate and attempt to enlighten these heavy-handed builders about the *value* of these personal commodities, the more difficult, demanding, and downright obstructive they will think that you are. And they may well be right. All of which will serve to ramp up your stress levels. Suddenly, the mother-in-law's offer of accommodation doesn't seem such a bad idea.

But, if you can't afford to vacate your premises for the duration of the work, then learn to cooperate, communicate, and tolerate your chosen builder and their operatives. Though, trust me, remaining in situ rarely makes for a happy camp.

So, apart from the cost and the major upheaval, are there any other downsides to improving your current home?

Well, not so many. Oh, yes, there are the neighbours. Have you told *them* about the noise that is going to be happening on a daily basis for the next however many months? There goes their cosy Pina Colada in the garden. What about the dust? The mud? The skip wagons, the stream of material deliveries, as well as the sweet birdsong (and occasional swear word) of the chirpy builder at work?

You see, if you've decided to stay put in your home for the long term, which is why you're doing the work in the first place, then you are going to want to keep those neighbours on board. After all, unless they're planning on moving any time soon, you're going to be stuck with them long after this construction upheaval has ended. So, be polite, explain the situation to them, apologise profusely in advance, and ply them with treats. Send *them* away for a week's holiday (though nothing too extravagant of course), even if you've got to stay put to endure the temporary mess that you yourself have created.

Even though altering and extending your existing dwelling can be stressful, exhausting even, I say go for it. It often makes most sense for the space that you have and the place that you love.

You chose your home for a reason. If that reason is still valid, then by all means alter and extend your existing property. Get the right builder in to do the work, and don't forget the neighbours!

MAINTENANCE

The word *maintenance* comes from the French word 'maintenant', which means 'now'. What we do now in the present, we don't have to do later. That's the whole essence of maintenance right there.

Step outside of your front door. Look at the timber barge boards that adorn your gable, or the fascias and soffits that support your guttering. Are they in need of a lick of paint?

If they are, imagine what is going to happen over the course of the next couple of winters if you don't treat them. The paint is going to flake. Then, without that layer of protection, the timber is going to rot and – ultimately – you're going to have to replace the whole lot.

But, caught early, treated and maintained, you'll double or triple the lifespan of the item. A stitch in time saves nine.

Warren Buffet, the sage of Omaha, is one of the world's wisest and wealthiest men. He's reached a ripe old age, and aims to be around for a while longer yet. His advice? Get yourself checked out. If you have a dink, a chink in your armour, something that's not quite right, get it sorted. Immediately. A problem ignored only becomes a bigger problem later on.

The same is true for the maintenance of your building.

I'm sure that many of us take the opportunity to have our car serviced at the same time as we put it in for its annual MOT. Parts get replaced, and potential issues are dealt with at source, in advance. That's maintenance, on a private scale. We all do it. It's just common sense.

Then, on a corporate scale, we have what is known as Facilities Management.

I once worked for a company called Land Securities Trillium, with 3,000 employees and fancy head offices in the City of London.

Some whizz kid had come up with the bright idea to sell a programme of maintenance to a lot of government and corporate clients. These included the BBC, the Department for Work and Pensions, all of the DSS buildings and Job Centres, and even the DVLA's driving test centres.

All of these organisations held vast amounts of real estate, and none of them were remotely construction-minded. They didn't know where to begin when it came to adapting or maintaining their own infrastructure.

Solution? Outsource to a company that was geared up for the task. That was where we came in.

We sold them a 25-year programme of maintenance for each of their premises. We promised to paint their buildings every five years, replace the carpets every five years, change the ceiling tiles every seven years, and re-roof every 15 years, or words to that effect.

They lapped it up. I had about £500 million worth of maintenance works in my database; that's £20 million a year for a quarter of a century. Not bad bread and butter.

A team of building surveyors would review the requirements for these works as they became due. Paint looks okay; push that back to year seven. Roof looks a bit knackered; bring it forward to year 12. Savings were offered for work deferred and monies reclaimed for work brought forward.

The clients understood that they had to maintain their properties, and we were the lucky recipients charged with attending to their needs.

We were also tenants in one of their buildings, taking up a whole floor of one of their units. We employed a subcontractor to do the painting. One undercoat and two coats on top.

Guess what. When they came to paint our floor of the building, they only did one undercoat and one top coat. They were cutting corners. Basically having us off.

Well, an eagle-eyed employee spotted their deliberate mistake. We pulled them up on it. Cue a rebate. No doubt they carried out this ruse as standard. We could have taken them to the cleaners – a major boo-boo on their part – and halved their remuneration.

The lesson to take from this is, when you're painting your own client's premises or your own boss's house, that's the one you make sure you get right!

At the other end of the scale, I once worked with a brilliant craftsman on a major conservation and restoration project in London. When that was complete, after three years of endeavour, this bloke's next appointment was as the resident maintenance man at Westminster Abbey. He had carte blanche to do whatever work he felt was most needed on a day-to-day basis. Like an artist, touching up his own canvas.

He wandered around the place, fixing things, sprucing it up, and securing this national asset for many years to come.

Someone had obviously recognised the importance of not leaving these precious things to simply fall into disrepair. They had a budget that afforded the man a wage, and I bet it was a fraction of the cost of rectifying the eventual disrepair and ultimate failure of this cherished building.

The Abbey, by the way, is adjacent to the Houses of Parliament; one can only wonder what might have been if a robust programme of maintenance had been in place there in years gone by.

Like the old saying goes, a penny now or a pound later.

That is the power of maintenance.

9. TIMELINE OF A PROJECT

PLANNING AND BUILDING REGULATIONS

Fail to plan. Plan to fail.

As well as being a surveyor, I'm also a writer. Before I embark upon any creative project, there's a certain phase, probably my favourite part of the whole enterprise, where I just sit and think. Stare out the window, go for a walk, let my thoughts meander, and then hopefully converge. From that daydream, a plan emerges.

The same should be true for your building project. Before you commence proceedings, before you put a spade in the ground, let it play out in your mind's eye. What exactly are you planning to do? How is it going to look upon completion? What are the different stages you need to go through in order to achieve your goal?

For a building scheme, you need to think about practical measures. What time of year is the work scheduled to go ahead? If you're re-roofing your house, is winter really the best time? I'd suggest not. For one thing, no one likes working in the wind, snow, and rain. And your roof protects any property from these elements; you are going to be left seriously exposed if you remove that all-important cover when the weather is at its worst.

You want to do that kind of work in the spring, summer, or autumn, right? Say it's the spring you're aiming for. You need to get quotes, and you need time to then evaluate those quotes. Plus, the better builders are booked up months in advance, so if you want to do your roof in, say, April, you better pick up the phone to those contractors at the start of January or by early Feb at the very latest. It's called having a plan.

I remember a personal fitness trainer once telling me he'd just had a call from some prospective clients. A married couple. They had a forthcoming holiday and wanted that beach bod to show off when they got there.

Neither of these potential customers were in particularly good shape – exercise just wasn't on their agenda – but now they thought they should pay it some mind.

'When are you going away?' the fitness instructor asked.

'In two weeks' time,' came the reply.

The exasperated trainer hung up the phone. You can see his point. Not enough notice; not enough time.

Give yourself the space to think about what you want to achieve. It might look like you're staring out the window and not doing very much, but this time costs you nothing, and it could save you a fortune.

1. What are you trying to do?
2. How are you going to go about it?
3. What are the steps – along the way – that you need to take?

Be the man, woman, (or whoever), with the plan!

I've seen so many projects flounder through indecision.

Your builders need direction. You can't just make it up on the spot or change your mind on a whim. At the end of the day, or at the end of the week, those builders want paying. They can't go home on a Friday with half their wages because you weren't sure whether you wanted something placed on the left-hand side of the room or on the right-hand side.

Know what you want to achieve before you begin. That's your plan. And then stick to it unless you meet a curve ball and have to get creative. But that's a different story. Make decisions, and know what you want.

You'll also have to think about how your construction project impacts its surroundings, and what permissions you might need to obtain before you proceed.

Planning regulations are a bit of a nightmare. Or maybe they are a complete nightmare; I'm not sure.

Your project might be delayed for months or even years, depending on the scale and complexity, once your application disappears into the system.

On *Homes Under The Hammer* recently, we met a guy who bought a derelict property and was going to do it up, bring it back into use, and add value where none presently resided.

When the TV crew and presenter turned up a good while later, they found the project stagnant.

The developer was an experienced builder, a salt-of-the-earth older chap who'd been around the block a few times, and when the host of the show expressed surprise that a seemingly straightforward enterprise was stuck in the mud, the bloke said he was still waiting to hear back from the planning department about his proposal.

With far more diplomacy than I could ever muster, he said, 'I guess they must be busy.'

And that's the planning black hole right there in a nutshell.

While there is clarity about what is permitted development, e.g., the footprint of an extension to an existing property (three metres farther out, and no wider than the present width of your building), everything else seems to be subjective.

Some planning officers might come on board and endorse your vision and enthusiasm, whilst others look to say 'no' at every opportunity for fear of approving something that might come back and bite them on the bum.

I have a friend who owns an old warehouse down by the docks. Probably two or three hundred years old. Eight storeys tall, he uses it as a rehearsal studio, and the roof is absolutely knackered. He wanted to re-roof it and actually wanted to make a feature of the new roof. He was going to make it a statement piece.

The boats and cruise ships coming up the river to the landing stages would be treated to a light spectacular.

My friend put in the required planning application, and the planning officer duly came around. He noticed the age of the building and – up in the loft space – he couldn't help but notice and admire the rusting, crumbling original iron trusses and girders.

'You'll have to preserve those,' he said. 'Historical importance.'

Now, my mate doesn't use the top floor/loft space of his building for the simple reason that the slate roof has perished, and pigeons have free access to it, as does the rain and the snow.

All he wanted to do was put it right. To have a fit-for-purpose weatherproof roof and maybe even make a bit of a feature/showpiece out of it, too. Of cultural importance. A gift to the city. At his own expense.

But no. It's apparently a historical feature.

'Yes, but it's completely knackered,' my mate pointed out.

'Doesn't matter. It has to remain.'

Eventual solution? My dear friend has now put a new roof over the *existing* rusting metal framework.

No one will ever see it. But it's there.

And if someone should ever wander up there, and be unfortunate enough to be struck on the head by a falling rotten iron beam from the original decaying structure, then we can call it *Death by Preservation*.

Let's see what the planning department has to say about it then.

Building regulations are also a bit of a minefield. While planning laws relate to permitted development and ultimately green light or refuse any scheme, the building regs state the standards that must be achieved once you go live. They concern themselves with the specification of your materials, the quality, the energy efficiency, and the sound performance and the combustibility of the products that you choose to use.

It's best to get a professional on board to sort out the application and any permissions that may be required. The Building Control Officer will need to be consulted anyway to check your construction meets the regs.

An architect can look after you in all of this regard, of course, whilst other construction professionals are also available.

Now I'll tell you about a distant relative of mine. He once built an extension at the back of his house. It was a self-contained granny flat. He just built it and told no one about it. Once something is built, and has been up for a period of 10 years, it's deemed to be a permanent structure. It's there to this day. He got away with it.

Author's note. I'm sure it was a 25-year rule when I first studied construction three decades ago. Now, it appears to be ten years, although in rare instances, if no objections are subsequently raised, it can be as little as four years. Consistent or what?!

Anyway, with other developments, whenever my relation has had to apply for planning permission – and where legal notices have been placed on nearby lampposts to alert local residents and give them the opportunity to raise objections – he has simply gone for a walk and torn those notices down. No one was any the wiser. With no objections, he was free to proceed. And, like I say, that's just for the things that he has cared to share with the planners. Sometimes, he's just gone straight ahead.

Still, sticking a notice on a lamppost hardly seems like the most efficient way to advertise and notify the public of proposed developments, so it's no surprise that people get away with things.

And when you hear some of the nonsense that some of these officials come out with, like my mate with the rusting trusses in his old warehouse, you can hardly blame my relative for taking things into his own hands.

Of course, I'm not advocating that approach! I'm just going to wish you luck with your project, and hope that you sail swiftly and smoothly through the planning laws and building regulations as you go.

Build without approval at your peril. Be prepared to have to knock it all down if you get caught out. You'll know in your heart, or in your gut, if you're flouting the law to an unacceptable degree. Whilst I do believe the opinions of planning officers are subjective, there are also clear lines in the sand that shouldn't be crossed.

In 2023, in a town called Kilbirnie in Ayrshire, a mansion belonging to a local businessman was ordered to be torn down as it had been built without planning permission. The local authority said the property had a negative impact on the area, did not fit the 'established character' of the surrounding countryside and, to make matters even worse, had been constructed in a place of previous shallow mining, meaning it was at risk of collapse. In other words, they could not have got it more wrong. Retrospective planning application was denied, as was the appeal to the Scottish government, who upheld the local authority's decision. Now the five-bed mansion has to be demolished… gone… wiped from the face of Google earth.

So, ignore these important, though often frustrating, rules at your peril.

The time required to satisfy the legal conditions attached to your project needs to be factored into your overall construction programme. The recruitment of qualified professionals to help you navigate these hurdles should form part of the master plan that you dreamt up while you were staring out the window when you first formulated your idea. But then you all knew that anyway, didn't you, because you've been listening.

I'll say it again. Fail to plan. Plan to fail.

PROGRAMMING AND PROCUREMENT

Your programme is the step-by-step breakdown of your overall plan. You need to break your project down into bite-sized chunks or individual pieces from the start to the finish. You then want all of these individual activities to flow with the grace of a Mozart concerto. Each item of every phase organically achieved, in sequence, so that – by the end of things – you will have succeeded in your goal. And a big 'well done' to you, too.

Say you want to demolish an existing bungalow and build a new two-storey family home on the same plot. Hopefully, you've purchased the property before you attempt to knock it down, but assuming you have, and you have received all of the necessary permissions, you can begin.

The first stage, as stated in the previous chapter, is to stare out of an available window and make your plan. Note your thinking on paper or a spreadsheet. Now you're ready to drill right down into the itsy-bitsy details.

If you're the project manager, these details are the things that you'll need to consider. If you're appointing a main contractor, these are the items of work that they will need to interrogate and include in order to meet your brief.

First step? Well, you can't just begin. You need to establish a presence onsite. I suggest you are going to need to erect hoardings around the site/work area. This will protect the public from any flying debris, and also keep the site secure from inquisitive and mischievous children, ruthless passing scrap dealers, and opportunist or professional thieves when you start to bring valuable materials and products onto the job.

You need to complete your site set-up. Do you need site cabins or welfare units (e.g., toilets, canteen, and a changing room for the workies)? You are going to need to be self-contained for the duration of the project.

Your main cabin might have a small meeting area where clients, contractors, architects, and other interested parties can get around a table and sort things out in weekly, fortnightly, or monthly briefings.

Long gone are the days when a single portaloo would suffice for the builders onsite. Especially for projects that are of significant cost or duration.

So, step one is to establish yourself onsite. Make the site secure and provide the necessary habitation for the professionals, contractors, and subcontractors involved so that they can operate safely and (hopefully) happily.

Step two for this particular scheme is the demolition. What plant do you need to carry out the work? You need to isolate existing services; you don't want anyone cutting through a live electric, gas, or water main.

Are you going to use a bunch of people armed with sledgehammers, or will you hire a JCB and a qualified driver for that machine to tear the existing roof and external walls down?

Does the JCB have one of those menacing steel drill bits on the front that can chew through the concrete foundations? Can you then change that bit for a bucket to scoop up the remains and dump them in a skip?

Allow the time to demolish the existing property and dispose of the resulting debris offsite, to a suitable place. I don't want any fly tipping, thank you very much. Remember to include these costs in your budget.

Now you have a vacant site and your plan is starting to take shape.

Who's doing the setting out? These are the calculations that make sure your development is situated in exactly the right spot and correctly orientated. Do you have an engineer on board to do this?

A good site manager or project manager may well have the practical skills; just make sure you position the new building in the right place. You don't want to get too close to the boundary fence or a neighbouring property. Please take the time to get this right. Sleep on it, by all means. Check, check, and check again.

In 2014, a new cinema being constructed in Cambridgeshire was re-built *twice* when it was found to be positioned incorrectly. That's a costly mistake. People blame GPS and they blame the architects. I think it's worth taking the time to go over everything before you commence with the construction activity.

What's the difference between a chicken and an egg?

Answer. The egg is involved. The chicken is committed.

And you are about to commit. Big time.

So, make sure you know *exactly* where your building has to go. I mean, to the millimetre. Once you are satisfied (and you have the agreement and backing of the professionals around you), then you can begin.

Step three, dig the foundations for your new construction scheme.

Step four, put up the walls.

Step five, put in the intermediate floor joists and the roof trusses up top.

Then, it's the turn of the roof coverings.

Install the windows and doors. Now you have a watertight building.

Internal walls, electrics, plumbing, plastering, sanitary ware, kitchen, fixtures and fittings, and decorations.

Do it step by step. Hopefully, one thing following seamlessly after the other. Like a Mozart concerto.

Otherwise, if there is a lag in the programme, you may have ongoing costs accrue more than expected, such as your scaffold, site cabins, site manager, and plant, etc. The meter will still be running on these items while your project is at a standstill for whatever reason. One common cause for downtime is a lack of resources. You've not managed to find an electrical contractor or a joiner or plasterer. No one is willing to work for the budget you have in mind. Or they've heard you're a hard taskmaster, or a poor, late, or even a non-existent payer. Therefore, no one is coming.

But, assuming you're okay in that regard, you also have to think about the next item in conjunction with your programme. And that area is procurement.

Basically, procurement is purchasing. Think of your project as a baby that needs nurturing. An infant that needs feeding.

You have to provide whatever the baby (your project) needs as and when it is required. It's easier said than done. It's a bit of an art form, in actual fact, but with a smidge of aforethought, planning, and communication (and often a bit of research and asking around), it is all perfectly possible.

When it comes to purchasing, you need to be one, two, or even three steps ahead of the game. You need to be a bit of a chess master, to be perfectly honest.

Just say that you or your main contractor has sent out some tender enquiries for brickwork subcontractors.

You're supplying the bricks. Are the ones you want available? Get on the phone. If you find someone who has some in stock, at a price that you like, then lock them down. Of course, you can keep phoning around to see if someone else has the same product in the same quantity for a better price, but if not, you'll be back on the phone to the earlier guy, hoping that no one has snapped up the items you want and need in the interim.

And when the walls are finally up, and you have someone coming in to fit the roof trusses, have you secured the slates or tiles that you want to go on the roof? Is there a shortage of that product? Is there a lead-in time?

If your programme of work states that the roof coverings are to be fitted in four weeks' time, and you find out that they can't be got for eight weeks because there's none in the country and they're coming from Spain, then you suddenly have a four-week delay on your hands. That's a month's extra wages for the site manager, the project manager, the site cabins and, well, you get the idea!

Now, if you'd made that call four weeks earlier, then guess what? The materials that you wanted would be landing onsite right when you needed them.

A couple of times in my career, I've had clients say that they would do the procurement on our behalf. They had preferential rates with their suppliers, they said. They just wanted us to fit whatever they supplied.

And it usually doesn't work.

Builders generally have preferential rates with their suppliers anyway, so costs even out. But then you have the prospect of sending tradespeople to site and the materials haven't arrived. You end up chasing the client or main contractor, saying you've got people standing around with nothing (i.e., no materials) to fit.

On one project of mine, a delivery eventually turned up mid-afternoon, transporting a tiny amount of what we actually needed. I'd already spent most of the morning on the phone asking my client for these materials to be urgently sent. Within a couple of hours, the guys we'd sent to site had worked their way through the short delivery, and I was constantly on the phone (on top of the half dozen calls that I'd made earlier) telling my client to supply more of those materials. A case of 'too many cooks…'

Nowadays, I just add a couple of grand onto our quote for the standing around time. The client thinks that they're being clever and saving money by using their own suppliers, but (as mentioned) the rates the builder gets probably match that of their clients, because these contractors are doing this day in and day out, spending money at the merchants. So, now, my clients get to pay extra for the grief that I'm pretty sure they are going to cause me.

Do your own procurement, or leave it to the builder. Just do what is needed to properly resource your job.

Procurement is massively important. Plug those gaps. You're trying to create a sequence of events within your programme, and each of those work events needs to be adequately funded, resourced, fed and fulfilled.

After all, it's your baby.

CHOOSING THE RIGHT CONTRACT

What is a contract? I'm sure most of us would agree that it is the written record of an agreement between two people or parties to do something or to provide goods or services for an agreed sum. That is correct.

What *constitutes* a contract? What *constitutes* a construction contract?

The first thing to say is that a handshake can be a contract. It's not the most robust, provable, or defensible arrangement in the world, but it can still constitute a legal agreement between two people.

As far as I'm concerned, the least distance travelled from that simple handshake is the best form of contract that you can make for your project.

I studied contract law at university as part of my surveying course. It was by far my favourite subject, and I even got distinctions for it. I think it spoke to my core sensibilities.

I hate to see 50 or 100 or 500-page contract documents attached to a project. What the hell are they hiding in there? What on earth are you signing up to?

The clauses contained in voluminous contracts can cover every conceivable eventuality, from the outbreak of war to the end of the world. After which, they then still hope to sue you for non-completion of the works!

Jake Eberts had an essential role in getting the movie *Dances with Wolves* funded and made, and did it on a two-page contract. He had nothing to hide and shared the risk and reward with the filmmakers. So why should a building project that costs maybe fifty or a hundred grand, or even a million quid, need a 50-page contract? It beats me.

That Kevin Costner movie took in about a billion dollars – on a two-page contract – so a scheme costing a mere fraction of those figures does not need a contract that's twenty times the size.

I work for a company who have their own legal department. They are so risk-averse it's a wonder that we're ever allowed to set foot onsite. And our clients are generally of the same mindset. It's like watching a game of tennis as the respective lawyers serve their terms and conditions. Volley back to you with a few changes, now back to you with our amendments of the same.

Can't accept those T's and C's. He's our rejoinder.

Here's ours.

And here's ours.

Meanwhile, the job has already started onsite and may even have been completed in the interim!

Now we can't get paid because we haven't signed the contract. But we can't sign the contract because the conditions aren't acceptable. Can we please just sign it anyway in order to get paid?

If it ever goes to court because we signed up to some onerous clause within the contract, I'm sure we would mount a defence at that point. After all, we have a team of legal experts at our disposal.

But we all seem to want to talk our jobs into a legal quagmire. The language employed in these contracts is often open to interpretation anyway, so I don't know why we're all getting so hung up on it, especially at the expense of doing some actual work.

So, what does this all mean for you?

Say you're getting some work done on your house. You've found a builder who's willing and able to undertake the task and you've agreed a price with them.

There's nothing wrong with you knocking up a simple one or two-page document outlining what is meant to happen from thereon in and how much it is going to cost you. In fact, it's a must.

Imagine if it all went wrong. Imagine that you're down the pub afterwards bemoaning your lot to a friend... the conversation would probably go something like this.

'So and so was meant to turn up and do my loft conversion. Put a new dormer in. New en-suite. Plaster the new rooms and put new lighting and heating in. I'd agreed a price with him, fifty grand. The bugger's gone and let me down. I gave him a ten grand deposit, thirty grand a few weeks later, he's only done half of the bleedin' job and now I can't get him back to site to finish it off, etc, etc.'

That, right there, was your contract. Write it on a page. Loft conversion consisting of blah, blah, blah for a total sum of £50K plus VAT. Get the contractor to sign it. You countersign it. Each of you keeps a copy. You know what you're getting and for how much. The builder knows exactly what he's expected to provide and how much he will be paid for doing it. Rocket science, it ain't.

If it all goes to pot, and you end up in a small claims court or wherever, someone of note will see that document as the Holy Grail of this particular dispute. Think Judges Judy or Rinder. Otherwise, it's just one person's word against another's.

I remember seeing a programme once about the rise of Microsoft. They'd been charged with some offence regarding the monopolisation of their business, and it was alleged they had attempted to crush opposition in the marketplace in a blatant disregard of consumer rights.

This was all back in the early days of emails and personal computers.

When the esteemed judge asked Bill Gates which person had sent one particularly incriminating email, which effectively stated 'crush all opposition', Gates replied with a smirk, 'the computer'.

He was challenging the judge's intellect, which is never a good idea.

Microsoft lost the case.

You see, regardless of the number of pages and clauses your contract has, it may still come down to the interpretation of an arbitrator. They will weigh up your conduct, the spirit of the contract, the communication that took place between both parties and the actions that were taken thereafter.

As Gandhi once said, if you are in the right, nine times out of ten, the law will come to your aid.

For contractors and for many construction projects, the choice of contract generally comes down to those offered by the JCT or the NEC. The JCT is the Joint Contracts Tribunal. Their suite of contracts includes minor works, intermediate, and major works options. They have a design and build version, and also something called the Standard Building Contract, which is recommended for large and complex projects.

Why call something *standard* when it is aimed at schemes that are anything but? That's the legal conundrum right there, in a nutshell. I think they do these things to keep us ordinary folk in the dark. That's why they can charge the fees they do, because they speak a foreign language called legalese.

Nowadays, many construction projects are done under the NEC library of contracts. This is the National Engineering Contract group. It used to be the preserve of engineering and infrastructure.

Increasingly, they are used in construction, though I can't for the life of me think why. Engineering and construction are cousins; they aren't the same thing. Still, no one seems to bat an eyelid these days when they are offered up as a route to contract for construction.

So, for large building projects, you're generally looking at either the JCT or the NEC option. For your own private development, you can always draft your own version of a legal document. The laws of this country ask for four things to be in place to form a contract between two parties.

The first is an *offer.* Someone asks someone to do something. Will you do this… etc. Build me a house, extension, garage, a stable for the horses, or tarmac my drive. That's part one.

Then, someone accepts that offer. A clear communication that the person, company, or entity agrees to do the thing that was put to them in the form of an offer. *Acceptance* is the second tenet of a contract.

There must also be something called *consideration.* This means that something of value is being exchanged. It doesn't have to be a significant sum or even an equal amount to what the other party is offering, but it has to be something. If someone says they will build you a new house for nothing, then there's no legal contract. Yet if you offer them a pound, and they are daft enough to accept it, then that counts as consideration.

You've probably heard the expression 'Peppercorn Rent'. Sometimes, a landlord might put someone in a property for a reason that makes sense to them (maybe security or whatever). That person would have no rights if they didn't contribute something to the bargain, so a single peppercorn will suffice. At least it's *something* under the eyes of the law.

There are many ways to enter into a contract. The final legal issue for making such a thing binding is *intention.* Were you just chatting drunkenly in the pub, or did you actually intend to build so and so's house for a certain amount of money? Did you mean for this to be a real and proper job?

Offer, acceptance, consideration, and intention. Draw that up on any napkin you like. It's binding.

And don't forget to get both parties to sign it. Job done. That's a contract.

But beware what you sign up to. If you sign up to do something, you're pretty much bound to deliver. Most judges and arbitrators will take your signature as your bond, with no going back.

I once heard about a case where a private client asked a builder to do some work on their house.

The inexperienced client asked for something that the builder *knew* would not work. It would almost certainly fail. I think it was a kind of roof tile on the house or a roofing slate on a roof with too low a pitch.

Oh, but the client really wanted that slate roof. (They don't work on roofs below a 30-degree pitch.)

The builder pointed this out. Said it would fail and that he wouldn't do it.

But the client persisted. They thought it would look good and really, really wanted it.

The builder said no again.

She brought out the cash. Do it anyway, she insisted.

The builder saw the pound notes, then did as the client instructed. The roof subsequently failed and leaked.

And she sued him.

And won.

In court, though she clearly admitted that she had instructed – almost demanded – that the builder did as she asked, she said that she had not fully understood the consequences of her request.

The builder, as the expert, should have continued to refuse to do the work, as they had initially done. The builder, even under instruction, should not have done something that they knew to be wrong.

He signed on the bottom line to deliver a disaster, and the judge took a dim view of it.

Let your signature be your bond. Understand the responsibility you take on when you become party to a contract. Keep it as simple as possible, and deliver your side of the bargain. The law will expect no less.

Choose your contract and your clients carefully, and don't sign up to anything you don't understand or agree with. Or to something that your conscience and experience tell you is not going to work.

That's how to choose the right contract.

CHOOSING THE RIGHT CONTRACTOR

How do you choose the right contractor for your construction project? Well, the first thing to say is that the importance of making the correct choice *cannot be overstated.* It is absolutely vital to the successful outcome of your scheme.

If you are a betting man or woman, then you will probably have heard the expression *horses for courses.* This means that certain racehorses are better suited to certain racecourses. Some tracks are left-handed, some are right-handed, some are straight, some are in the round. Some are excessively undulating, while others are flat. All of these characteristics generally appeal to some horses more than others. That's why some horses perform better on certain courses than they do elsewhere.

I would say that the same is true for your project. Some contractors will be better suited to one type of work than others. So how do you choose between them?

Experience is key.

I was once asked to tender on behalf of my employer, a building company, for a new access road, parking areas, and footpaths on the grounds of an existing pharmaceutical factory.

I did my site visit. It looked like a nice job, they were a good client, and they appeared to have plenty of cash in the bank.

And guess what? I declined to even quote for the work.

Why not?

We had no experience in that field.

I would have had to subcontract out all of the different elements. The groundworks, the setting out, the tarmacking, the flagging, the bollards, the marking out of the parking bays, and more. We did not have in-house labour to carry out any of the work, and would not have even been able to effectively supervise or sign off the work in terms of quality as we simply did not understand the requirements of a project of this nature.

It was best left to the people who did this kind of stuff day in and day out.

You'd do well to remember that when you are looking for someone to deliver your own scheme.

Whether you are a client looking for a main contractor, or a main contractor looking to subcontract work packages, or even if you are a private homeowner looking to get a builder in, the process should look similar in every scenario.

You want to get three quotes from potential suppliers if possible. You want all three to be reputable.

If you have prior experience with these people, that's a bonus. If you have to go to the market, ask around and see if anyone can recommend someone to you.

Then, do a bit of fishing. Check out their website; look for reviews online. You may even find news items concerning said party. Take note of anything that comes up, good and bad.

When you pick up the phone or send out your tender package, you may find that some contractors are simply too busy to fit you in. If possible, keep looking until you get your three quotes.

You want to work with people who are a good fit for your scheme. You'll want people who have experience of this type of project.

Ask for examples. Go visit these companies' previous or ongoing projects if at all possible.

If you are looking to get your loft converted at home, you don't want to employ someone who has never done one before. You don't want them learning on *your* job. Every building scheme is unique in its own way. Each construction site or individual property presents its own challenges in terms of access, location, conditions, and restrictions.

An experienced contractor will have met many of these challenges before if their bread-and-butter work is in line with your own requirements. That's the sort of contractor you want to appoint.

You're looking for quality assurance, availability, a reasonable price, and good communication and understanding between yourself and the building company. Therein lies the path to success.

You don't want to be in the dark about who these people are, who they've worked for previously, and how satisfied those customers were (or weren't). Caveat emptor. Let the buyer beware.

Do your due diligence before you award the work to someone. Otherwise, you could be in for a world of pain.

Personal recommendation is usually a good unit of measurement. There's nothing better than word of mouth when it comes to promoting anything. In turn, I would say that experience is the one fundamental that can't be bettered.

Get a plasterer to do your plastering. A roofer to do your roofing. A plumber to do your heating. An accredited spark to do your electrics.

And, overall, a small builder who does extensions to do your extension, and someone who does new-builds to build your new house.

But, at the risk of repeating myself, try to get three quotes for whatever it is that you need. It's called doing your homework. Covering your arse. Things can still go wrong, but at least you will have given yourself a fighting chance. That's all you can do and all you can ask of yourself.

And remember that you're the boss and that the customer is always correct. Choose well and be understanding, supportive, and appreciative of a good builder. By which I mostly mean pay them!

Do all of the above and you're on the right path. That's how to choose the right contractor.

Good luck.

ON TIME AND ON BUDGET

How hard can it really be to have your construction scheme come in on time and on budget?

Well, suffice it to say, I think that to achieve both is virtually the Holy Grail of construction.

Of course, you could always finish *ahead* of time and *under* your budget; however, just getting those two things right will do us very nicely indeed, thank you.

If you do try to finish early, you might put yourself and the entire project team under unnecessary pressure. What is the point?

Remember the Titanic? All she had to do was sail across the Atlantic and arrive in New York in one piece. Well, some bright spark had the idea of asking the captain to speed up so that they might get to their destination early. 'Imagine the headlines' was the motivating message. Well, they certainly made headlines, all right. Just not the ones they wanted.

On time is perfectly fine.

So, why do so many projects fail to meet their deadlines?

Weather could be a factor, but you can and should build this into your programme. Availability of resources such as labour or materials is another factor. Again, you can always include a margin for error (called 'float') in your initial calculations.

Don't put yourself under unnecessary pressure from the outset by overpromising something that you can't deliver. It is better to have realistic expectations about your timescale from the get-go.

You need to plot and plan each and every step of the project, allow enough time for each step to then happen, and give yourself some breathing space. And also allow for a little buggeration.

If the client wants the work done in ten weeks, tell them it will probably take 12 to 14.

Make sure you get the start date right. If you are finishing another job and you can't start for four weeks, then don't promise them that you'll be there in two. Otherwise, you're going to be two weeks behind before you've even begun.

You want a good, organised flow of labour and materials throughout the project to give yourself a chance of coming in on programme.

Someone needs to be in charge, like a good project manager.

Duncan Bannatyne on *Dragon's Den* once said that whenever he is getting something built – like a new gym or hotel – he ties his chosen contractor down to a fixed price and programme period, with liquidated damages for late delivery.

So, if the building isn't finished on time, the contractor has to pay, say, twenty grand a week in damages. This will compensate Duncan for the loss of profit on the venture in the weeks that his new business can't open. It also serves to focus the minds (and available resources) of the contractor.

The same with the price. Get it agreed upfront. If anything varies on the scope of works, these items are costed separately and added (or subtracted if it's an omission) from the contract sum.

But, generally, you want agreed costs before your project begins, and you want the contractor to deliver for that price.

As the client, you need to be strong with your contractor. I've heard every excuse in the book when it comes to people asking for a bit of extra cash.

Prices have gone up. It's proving more difficult than first anticipated. I've made a mistake with the price and I can't possibly do it for that amount.

Now, I'm a firm believer in paying people well. It serves as great motivation. It means that people prioritise your work and come running when you call. Still, you can't immediately roll over when *the price was the price*. Sometimes you just have to say, 'Tough. Get on with it'.

Help people out, by all means, but don't be a soft touch because you will get asked for a lift, trust me. Listen, and then explain, whether that is in the positive or the negative in response to their cries.

We all have our crosses to bear. Don't carry anyone else's.

In my experience, I would say that maybe one in ten construction projects come in either under time or under budget. Then maybe three in ten actually come in on time and on budget.

Six out of ten, therefore, fail on either one or both of these counts.

Apart from the weather, a lack of resources, or the outbreak of war in some part of the world with its knock-on effects, what is causing over half of construction projects to fail on these fundamental points?

I think the major cause is simply poor performance. This relates to the planning, procurement, and overall project management of the scheme.

It's a sad state of affairs to think that this is almost the norm. By my reckoning, over half of all construction work is remiss in this regard.

Construction work takes place, and is taking place right now, on almost every street and in every town and city in Britain. With such experience, and such prevalence, why is it that we can't get these things right the majority of the time?

Well, let's not kid ourselves; builders aren't rocket scientists. They tend to be blokes with fat fingers, doing as little work as they can for as much money as possible. That's the grim reality.

Money talks and it is a major motivating factor when it comes to attracting the best of the bunch, but we are still only dealing with and working with builders at the end of the day.

Also, projects are live. They are real. Building sites are often busy and dangerous environments. People can get hurt. In a recent survey by the Health and Safety Executive, 40% of the construction workers interviewed said they believed that accidents were 'inevitable' in the building industry. They've seen the chaos that can exist onsite, the clamour to get the work done if you're on a price. Sod anyone and anything that gets in the way. People throw caution to the wind when they're chasing a buck or they're forced to meet a deadline.

In other words, it can be quite challenging – what we do – and it's not always easy.

Building a house, or a stadium, or a new hospital, or a nuclear power plant, well, it's no walk in the park, I'll tell you that.

It's also generally highly visible. You can see the result of our industry's endeavours. And, because you can see it, you can easily comment on it even if you're no expert.

We are judged by our efforts, and that brings added pressure to the work that we do in construction.

And then there are the clients. Generally, they're no experts themselves, but they are the ultimate end-users, and (think they) know what they want. The builder has to interpret the demands of these people and deliver them something that they'll ultimately hopefully accept.

So, it can be difficult, but it can also be rewarding and occasionally even profitable.

Aim to finish on time and on budget. That is construction success.

10. HOW MUCH WILL IT COST?

FUNDING / FINANCING

I remember an old manager of mine who – whenever we received a tender invitation from a new client, (i.e., someone that we hadn't worked with previously) – would always ask me one simple question.

Have they got any money?

Quite important, that one.

Of course, most companies now have the ability to credit check potential business partners, and it's a very worthwhile exercise. If the person or the company making the enquiry to you has zero credit, and maybe some county court judgements against them, plus no floating capital, then how on earth do you expect to get paid when you've carried out construction work on their behalf?

I once received an enquiry from a contractor who was building a number of new houses in the Midlands. They wanted us to quote for the roofing works and – at a guess – there must have been a thousand roofing companies between their building site and our base about 100 miles away.

Alarm bells started ringing.

When I mentioned the company's name to our material supplier, he told me to steer well clear. He'd heard very bad things about this outfit, and he also gave me the name of another contractor who had previously worked for them. I called him up, told him the issue, and he gave me the same warning.

Our in-house accountant then did the obligatory credit check on our prospective client. That came back with blazing red flags, too. When their company director called me up to ask when I would be returning our tender, I told him about my reservations.

He heard me out, and then beseeched me to take on the job for them. He tried to explain his own company's version of events, how it was all someone else's fault, the banks had let them down, the subbies they'd employed were all useless, etc. I said that I couldn't ignore three express warnings. I actually asked him, 'How many warnings do you think I should ignore?'

But he needed the work doing, and 20-odd houses was decent work, so I proposed the following.

We would rock up to site on a Monday with men and materials. We would invoice on a Tuesday (the following day) for work projected up to Friday of that week.

If he paid that invoice on the Friday, a money transfer straight into our bank account, then we would turn up on the following Monday and carry on. That way, we would never have more than a week's worth of labour and materials at risk.

He agreed to my terms.

It worked okay for a while, and then they didn't pay us one Friday. So we went elsewhere on the Monday.

The site manager then phoned me.

'Your men haven't turned up.'

'Yes, that's because you didn't pay us.'

'No,' he said, 'we'll pay you this coming Friday. We pay you a week in arrears.'

'That's not correct,' I said. 'If we waited until the following week, and *then* you didn't pay us, we would have done two week's work. I've said all along that one week was the furthest that we would put our necks on the line.'

You have to be tough. Know what you're dealing with and protect yourself accordingly.

That particular debt paid, we returned to site, and we actually did okay out of it. By covering our backs.

Funding for any scheme has to come from somewhere. You don't walk around your local Tesco store or Sainsbury's on a Friday evening, fill up your trolley, and then get to the checkout not knowing how you're going to pay for it.

The same is true (or should be) for any development scheme.

Who's paying for it? How are they funding it? Where is the financing coming from?

If you have fifty grand in the bank and want some work doing on your house, you get a few quotes in and, assuming they are within your budget, you give them the go-ahead and they can crack on.

And then you pay them accordingly. You're the funder, from your own private finances.

Schemes can also be publicly funded from the treasury's coffers, of course. Local authorities, public bodies, utility companies, broadband providers, massive conglomerates.

These are all funders and financiers and instructors of construction work, but you still have to ask yourself the question, can they afford to pay you? Do they have convoluted accounting systems, offshore email addresses, and myriad fronts to hide behind? This will cost you time and cause you immense frustration when you try to get a response as to why your invoice or invoices haven't yet been paid.

I was in a meeting several months ago. It was a commercial meeting to discuss client debt and our anticipated paths to recovery.

One client owed us £100K.

They had also asked us to do further work to the tune of an additional £500K.

Our regional manager wanted to continue, regardless of the fact that they couldn't pay their bills.

So I spoke up.

'Has anyone read today's newspapers?' I asked. 'It says that the receivers are going in there next week. We'll be lucky to get the £100K back that they already owe us. Please do not carry out any more work or purchase any more materials on their behalf.'

Was I listened to?

Our monthly debt tracker meeting now discusses the £600K that we are owed by that client.

Do they have any money? Bear that one in mind before you carry out any work for anybody.

Not many of us have a bottomless pot of cash. Those people who do have a bit or a lot of money generally got it for a reason, which means that they are savvy when it comes to financing.

Their own money will be ring-fenced. Using OPM – Other People's Money – is the modus operandi of the smart set.

Banks love to lend money to the people who don't really need it. There are probably existing assets that they can secure their loans against. Things like that.

Developers usually include the cost of borrowing the required funding when they come to consider the viability of any project, while some eager entrepreneurs come racing out of the starting gates – procuring subcontractors and putting them to work – while they still don't have funding in place. Living on dreams; stretching others to breaking point in the hope that it will all come good in the end.

Many companies are founded on debt. I read in the financial pages of the papers recently about a man said to be worth £1.3 billion, whose company posts debts of £300 million a year.

I can't profess to know a lot about big business or financing, but surely he has got to turn a profit at some point? Maybe I'm just naive. I know for a fact that the aforementioned businessman won't be accountable for any of his company's debts; his personal wealth will be secured for him whatever happens.

If you're a private entrepreneur hoping to build a small property empire, you could do worse than get your foot on the ladder, buy a derelict or run-down doer-upper, and then remortgage once the work is complete. Use the equity that you've built up to purchase your next place, and so on and so forth.

If you have the funds to realise your vision, go for it. If you are the contractor working for that funder, check that they have the finances to actually pay you and the inclination to do so.

Don't work for crooks or fare dodgers, and make sure you get paid.

As my old nan used to say, 'When the money goes, love flies out the window.'

It's as true in life as it is for your envisaged construction scheme.

VALUE ENGINEERING

The client has a vision and a budget, but the budget doesn't quite match the vision, so what do you do?

Value engineer.

This is where you put your thinking cap on and ask how you can get as much bang for your buck as possible.

You can haggle on price, of course, but you might still be faced with a shortfall. You can always look at the specification. Can you choose a cheaper product? Go for UPVC windows instead of aluminium surrounds, stainless steel taps instead of gold!

You can also reduce the scope of works. Do less than what you first thought about doing or hoped to achieve. That's a realistic solution.

So, if you don't like or can't afford the quotes you get for that double-height extension, you may need to go single-storey to realise that kitchen-diner while foregoing the additional bedroom upstairs.

You need to stick to your budget.

If you've ever been skint, then you'll know the feeling of walking around a supermarket with five, ten, or twenty pounds in your pocket and a list of essential items that you need to buy.

You like Heinz beans, but they're a premium product. The Crosse and Blackwell variety tastes just as good (in my opinion) at only half the price.

Still too much?

Try the no frills range, which is even less. You'll (sadly) taste the difference, but it will leave you with a few shillings to put towards something else. That's value engineering. Thinking about each and every penny.

Now, a lot of clients seem to want a premium product for the smallest price possible. Simple common sense says that this isn't going to work. Quality costs – it just does – and as the old saying goes, 'Buy cheap, pay twice'. Or, 'You can't make a silk purse out of a sow's ear,' as an even older saying goes.

There are still things that you can do to save on cost and work within your budget. Shop around. You might find a cheaper supplier or a contractor who needs the work and will cut their cloth (and quote) accordingly. You can reduce the scope (i.e., the size of your project) to bring it back within your budget. You can alter the specification of materials to find a more cost-effective solution.

The important thing to remember is to do all of your thinking, your chopping and your changing, *before* you get started onsite with your development.

You don't want to run out of money halfway through, and you don't want to waste the money that you do have by making things up when the project is already up and running. Your builder will still want paying, even if you change something and make him do it three different ways. He'll want paying three times!

What can I realistically get for my budget? What do I need to change if I can't afford everything that I want? What is the overall aim of the project? Which items are most important to me and, therefore, non-negotiable? The rest might be surplus to requirements.

These are the questions you really need to ask yourself as you try to gain the maximum value out of your overall project spend.

That's how you value-engineer.

ESTIMATING YOUR COSTS

Very few of us have an endless supply of cash with which to play about. We usually have a budget in mind whenever we contemplate a particular project. Let's say that you've been dreaming about building a man cave/office at the bottom of your garden. You have a little bit of money lying around with which to finally achieve your goal. At this point, you really need to consider if your budget is adequate for the envisaged works before you waste any more thought on the venture.

There are several ways to evaluate the costs involved. The simplest method is to use a square metre rate, based on the footprint of your development.

Take a tape measure out into your back garden, pace out the area where you want your building to go, and measure the length and width of the envisaged plot. Say it's 10 metres by 8 metres, that's $80m^2$.

An internet search will quickly give you guide prices for new-build work. The parameters will be quite broad, as a whole host of factors can influence the scheme, such as location (Central London, for example, will have higher prices than elsewhere in the country), access, specification, and the like, but you could take a mid-point of the prices on offer and just use that as a simple guide.

So, if new-build domestic construction work falls within a range of £2,000 to £3,000 per m^2, take £2,500 as your average, multiply it by the footprint of $80m^2$ and you get a budget estimate of £200K.

Then try explaining that one to your partner!

Say you can't justify spending more than £100K. Reduce the footprint. Hope to get a local builder to do it for the £2,000 per metre mark. If your man cave is now 10 metres by 5 metres, giving $50m^2$ at £2,000 per metre, then you can still afford to dream. You're back within your £100K budget.

That, obviously, is a very rough guide.

You could always put the job out to tender. Real-life quotes telling you how much people are going to charge to do the work, how much they will put pen to paper and do it for, are ultimately your best guide.

You could break your vision down into its component parts. How much for the groundworks, the brickwork, the roof, plumbing and electrics, and so on? This takes a bit of knowledge and is a task best left to the experts, such as quantity surveyors and cost consultants. You could then package out the works and project manage it yourself, even though you might not make a saving and you could also gain a lot of hassle in the meantime.

Prices for recent, similar projects are a good indicator of what costs you might be facing; however, it can be a bit awkward if you ask your neighbour how much they spent on their new roof or on their kitchen extension. Try asking a family member instead. They might also be a bit more forthcoming with you if you tell them that the reason you're asking is because you're thinking of having similar work done to your own property. After all, no one likes a nosy neighbour or even a nosy relative.

Ultimately, there are many ways to get to a ballpark figure, but it is a bit more involved if you want to produce a more accurate estimate, which is only ever an educated guess anyway.

What someone is prepared to quote for the work… that's the reality! That's how you estimate the cost. Get someone who's prepared to do the job to tell you how much they're prepared to do it for.

Otherwise, you're just guessing. Or estimating, as we call it in the building game.

TENDERING

Every company needs to have a pipeline of work. They need to secure contracts as they have staff, wages, and other overheads to pay in the form of offices, vehicles, phone lines, computers, and all the rest.

So, how do they win that vital workload to support their ongoing costs?

Well, they do a bit of schmoozing to get onto a bigger company's supply chain as an approved contractor. Then, when invited to quote for some work, they submit their tender in competition with similar companies, cross their fingers, and hope that they've done their best to win the job.

Competing builders will be supplied with a set of drawings prepared by the client's architect, and a schedule of work or a bill of the quantities that sets out the work involved, as well as a copy of the specification of the materials. There will be a tender return date, and woe betide you if you miss it.

There'll be sleepless nights as you try to remember if you've included for everything, not missed anything out, and still kept your price competitive against rival bidders.

Then, it's up to the client and their representatives/advisors to make the final decision.

You might expect that the lowest submitted tender always wins, but that is not the full story.

Someone might be *too* cheap. Some firms bid low and then claim high on the back of the award. 'We didn't include for that,' they will say. Then, they will fight you tooth and nail for the duration of the project to extract every additional penny that they can.

They might have underestimated something, an error that will come back to haunt them, even bankrupt them. It could be a genuine mistake or possibly a cynical ploy to exploit the customer.

If you're the client, and you get four quotes back, with three of them in a similar range while one is extraordinarily cheap, you best ask yourself *why*. And then ask them *why*.

Ultimately, you want the job to be worthwhile to the chosen contractor. If they are making little to no money on it, then you aren't really going to capture their attention. Yours will be what is known in the trade as a 'hospital' job. They're basically just killing time on your project until something better and more profitable comes along.

It hardly classes as motivation. For you, this particular scheme might be the realisation of a lifetime's ambition, or the fulfilment of a ten-year struggle to get this project off the ground.

The last thing you want, in that case, is someone on board for whom the scheme is on the pay-no-mind list, almost an afterthought, and nowhere near the top of their list of priorities.

The good thing about tendering is it tends to keep people honest, in the sense that they'll want to give you a fair price for the work; otherwise, someone will nip in before them and grab the prize.

The tendering system does have its drawbacks though, and is (and probably always will be) open to abuse.

A tender document can be opened early, and the sums therein broadcast to interested parties, who can then offer up their own quote before the deadline at a figure just below the previous lowest price. It happens. It goes on everywhere. And not just in the world of construction.

Russia and Qatar to host the World Cup, anyone?

You can bet your house that money changed hands.

In 2023, ten building firms in the UK were fined a total of £60 million after a three-year investigation by the Competition and Markets Authority (CMA). The construction companies were found guilty of having operated a pricing cartel in relation to 19 public and private sector contracts with a combined value of over £150 million. The contracts that had been awarded concerned construction activity at a magistrates court, a police station, and a police training facility. In other words, nothing is sacred.

So, when you get your quotes back, set a little time aside to analyse them properly. Read the details. What have they included? What have they excluded? Make sure you are comparing like for like.

Then, also consider the following. Their availability to do the work when you need it to be done. How about quality assurance? Anyone can talk a good game, so do these people habitually do a good job? Can they provide references or examples of recent, similar work. Also, previous relationships can come into play when deciding where to award a tender once the quotes have all been received. Often, it's better the devil you know, rather than taking a chance on someone new. There's comfort to be had if you know the boss will pick up the phone when you call and have to kick ass accordingly should any problems arise.

There are also other things for you to think about.

If the two lowest tenders are there or thereabouts when measured against your budget, but one firm can't start for two months due to their current workload and the other one who came in slightly higher is able to start in three to four weeks, (where time is crucial, as you're going into the winter months for a roofing job or whatever), then by all means pay that extra few quid to get the project started when you need it to happen.

Even better, phone the second tender company up and ask if they can match the lowest quote. Give them a bit of fluff. 'You were slightly higher, but the contract manager says you did a good job for him on the last project, so if you can match the lowest price we received, we will give you the job.'

It sometimes works. If the contractor can't move on their price, maybe pay them that little bit extra if it serves you well in other areas (e.g., in terms of start date). It's only a process. You aren't duty or even legally bound to accept the lowest offer. Other factors can always come into play, but price is always a big one, obviously.

Getting three quotes for the work is all well and good, but in the world of small domestic, residential opportunities, people are often hard-pressed to get even one contractor to turn up and look at the job. You might try Yell.com and Check-A-Trade, but as soon as you say that you're waiting on other quotes, you'll go into the 'difficult customer' box and your next phone call to them will go unanswered; your voicemail will be unreturned.

You're trying to put your job out to tender, but no one seems interested. They don't know you and they are probably busy anyway, so why would they bother?

What can you do?

Well, try to be reasonable and accurate when describing the work that needs doing. The contractor or tradesman might then turn up and walk the job and come back to you with a price later on. If you've established an initial connection with them, trust what you are hearing from them, and if you can stomach their quoted price (if and when it comes), then you've not done badly. You're in the game. Someone might actually turn up and do the work, too!

Tendering is all about attracting several different offers for your work so that these prices can then be evaluated.

I've been on some great tender lists. These clients are the lifeblood of your company. They give you work and allow you to pay your bills. They could be government contracts, a local authority, service providers, private companies, entrepreneurs, developers, or individual homeowners.

I know of some builders who don't even advertise; they have a constant stream of clients. Their reputation speaks for itself and potential customers are queueing up to get them on board.

The reality for the majority of building companies, though, is rather different. They tender. They win one in three, four, or five of their submitted tenders, and that's just about enough to keep them going.

If you are in the business of submitting tenders for a living, then it's a good idea to keep a register of the tenders you're supplying.

Find an hour, or a morning every three months or so, to review the list of the jobs that you've recently tried to win.

You might find that there's a certain prospective client or contractor whom you've priced a dozen jobs for lately with no luck whatsoever. Not a whiff of success in the air for any of your bids, and no feedback from them either.

Ask yourself *why*.

Even better, ask *them* why.

You might be getting used as a pricing exercise.

That company or client might give all of their, say, electrical work to a particular outfit. You don't have a chance in hell of ever being awarded the contract, but they need to show to their bosses, commercial managers, or clients that they are getting the three requisite quotes in before they award the work to you-know-who. All of your efforts, the hours that you've spent completing those cumbersome tender documents, will always have been for nought.

So, get on the phone to these charlatans. Ask them who won such and such a tender. And the next one. And the other half-dozen that you have just priced for them.

If the winning contractor is always the same outfit, then it's time to wake up and smell the coffee.

Tell them that you won't be pricing any more of their tenders. It's a complete waste of time. Put them on the spot. Ask them how much did you lose those tenders by? What was the price differential between tenders A, B, and C?

They'll probably say that they don't have the time to break the figures down for you. Well, there's your answer. They can't be bothered. Neither should you. Time to say goodbye.

Incidentally, I once worked for a company that was winning only one in ten of the tenders that they submitted to a particularly prestigious client. A ten percent success rate; ninety percent failure.

At that point, you really have to ask yourself what it is that you're doing wrong. Are your prices too high (that's pretty obvious)? If so, then why?

When I joined, I managed to win two out of every three tender submissions. From 10% to 66.6%. That's not bad. And, yes, I did have sleepless nights, wondering about the figures that I was about to present.

I didn't manage to ask for all of the money that I wanted, because I realised other companies might want or need the job more than us. I've never wanted to win a job and not be happy that I've won it, as there has to be some incentive and profit margin involved. Still, I knew I couldn't charge them the earth as we would be back to the one-in-ten success rate, which is really no success at all.

The process of tendering, overall, is a good one. It keeps companies competitive and supposedly honest. It ensures value for money. Three firms prepared to do the same job for slightly different costs, based on their workload, buying power, their need for the work, and the size of their company overheads.

When tendering as a contractor, remember that the price you submit has to work for you. The client will want everything for next to nothing. Your price is *your* price. Get it right. Submit. Then let the client decide.

Take as much as you can get away with. But *don't blow the opportunity.* And don't ever try to win a job at a price you're not happy with. That's the golden rule.

THE SPECIFICATION

Crystal chandeliers, anyone? Italian marble worktops? Your choice of materials can affect the price of your project. Significantly.

We're talking about the inherent quality here, your own personal preferences, and your ability to shop around and secure the best value for the goods you choose to buy and employ in your build.

You can always look online for pre-used goods if, say, you are fitting out a second home that you are going to let out. You can find used kitchens, bathrooms, all sorts of stuff. It need not even be second-hand. Sometimes, people purchase items that they don't want, that can't be returned for whatever reason. They'll sell them to you at a decent discount. Get down to your local auction.

You'll find furnishings, rolls of carpet, and all sorts. Bargains are there to be found.

But that's not always an option. Say that your project needs a certain type of brick, plasterboard, roof tile, or flooring item. You're going to have to fork out the going rate or choose a cheaper, though still acceptable, alternative product.

For example, laminate flooring is cheaper than a hardwood version, but you'll end up replacing a laminate product more frequently than a hard-wearing wood. The maintenance or the lifespan of any particular product needs to be considered when making your choices.

Think about the return that these products are going to give you. How long will they live/last? How much enjoyment will they give you or your tenants or end-user? Do they add value in ways that can't be measured in pounds, shillings, and pence? In other words, are they worth the additional expense?

If you are building high-end properties for sale to mega rich clients, then your prospective customers will be disappointed and may well turn up their noses if, upon closer inspection, the fixtures and fittings of said dwelling fall shy of their exalted expectations.

Still, you can always try to get value for your money by bulk ordering. The same spec, but now you're a valued customer, a commercial type of guy rather than just a one-time purchasing punter.

So, if yours is a development of, say, ten houses, then you'll get away with installing ten of the same type of kitchen, and the same for your garage doors, front doors, windows and everything else. When you come to write out that cheque to your eager and excited supplier, make sure that you ask them for that (nearly always available) discount.

You can also pay less and still get the spec that you want by grouping your desired items together. You're bulk buying, and the overall price that you pay at the till should come down accordingly.

If the product you want is still over your budget and out of your price range, look for alternatives. For example, UPVC windows and doors now come in all sorts of colours and effects. You can achieve great things by exploring the marketplace and seeing just what products are available.

Go shopping. Get yourself down to the manufacturer's showroom. Keep an open mind as you wander around; you might be surprised by what is on offer.

Think about your end-user. What is the potential return on your proposed spend (i.e., your budget)? Then, use that figure to set your ceiling price when you go hunting for bargains while shopping.

Can you scrimp a little elsewhere in the scope of works to ensure you get the kitchen or bathroom that you want? After all, those two specific areas are meant to be the deal breakers and clinchers when people are viewing properties with an intention to buy.

There are many factors to consider when you come to choose your materials. It's one of the first things that you will need to think about as you begin to dream, plan, and conceive your project.

It's not all about aesthetics and appearance. It's also about cost, for sure, but cost has constituent parts, as we've just seen. It's about durability, sturdiness, and value for money in the long run.

So, choose your specification wisely. Or pay that eventual cost.

11. ADVICE FOR THE HOMEOWNER

DON'T BUY ON THE DOORSTEP

Before the advent of those 'No Cold Callers' signs, it was quite possible for salesmen to go door to door to try to sell their wares. In the early 1980s, when the working classes were encouraged to buy their homes, and did so in their millions, double-glazing was all the rage. This was a genuine upgrade to your newly acquired property. Before the age of the internet and edge-of-town showrooms, manufacturers and installers simply took their merchandise and their services to people's doorsteps. It was considered a legitimate way to place a desirable product in front of those who were most likely to buy.

But that was then.

Nowadays, I would never advocate to anyone that they buy on the doorstep.

When there are so many avenues to advertise and market your products these days, anyone who circumvents those routes and takes shortcuts should be viewed with suspicion. And rightly so.

The best products and the best companies need little introduction or marketing. Word of mouth from satisfied customers does the majority of their talking for them.

In fact, these businesses are more likely to weigh *you* up and ask why they should work for you. Can you actually afford them, they'll want to know. Are you likely to be a difficult customer? Should they even bother to touch you with a barge pole?

I had a knock on my own door about three years ago.

'We're painting the front of your neighbour's house,' they said. 'I noticed that your front wall is looking tired. The stone copings could do with a lick of the white stuff. Eighty quid.'

Bargain, eh? And the wall did look tired. We were only renting the property, but it was our home. We had a great landlord, and he left us alone to get on with things, so why not spruce it up for £80?

'Go ahead,' I said.

The pebble-dash render to the front exterior elevation was in a similarly depressed state.

After they'd painted the wall, there came another knock at the door and another proposal.

'£250 all-in,' he said. 'Front of your house will be looking lovely.'

I could just picture my wife's delighted face when she came home from work.

'Okay then,' I said.

Two hours later, I was £250 lighter.

Two weeks later, the wall and the front of the house looked exactly the same as they had done before that knock or those knocks on my door.

Whatever paint they had used was certainly no quality product. It was probably little more than nine parts water and one part emulsion. They'd also done no rub-down and no prep work whatsoever.

So let that be a lesson to you (and me). Don't buy on the doorstep. Ever.

You see, you can't take it back. When they're gone, they're gone for good. You don't get a receipt, and you don't get a six-month (or even a one-month) guarantee for whatever they've supposedly done. They won't leave the phone number of their complaints department either.

Now, my mate's stepdad was an old Irish bloke who was well into his seventies. One day, a couple of chancers knocked at his door.

He answered.

It turned out that his visitors were Irish, too. Oh, the banter they enjoyed. They talked about the old country, had a good old craic, before his new best buddies pointed out the state of his front garden wall. Did he want them to take it down and re-build it, they asked.

'Why not,' he said.

'Five grand,' they said. In cash, of course.

And he agreed.

Well, they knocked that wall down good and proper. It took them about half an hour to kick and sledgehammer the thing to the ground. A fine mess it was, hundreds of bricks across his lawn.

Then they knocked on his door again. They'd done the hard part, they said. Now, they needed paying so that they could go and buy the materials to build the thing back up again.

He gave them the five grand.

And they never came back.

So don't buy on the doorstep.

Now I'm Irish-stock myself, so I can say what I see; no wider aspersions meant or made here. Just don't buy on the feckin' doorstep!

A friend of a friend recently inherited a house from her dear old mother. The place was tired – stuck in the seventies or eighties – and she had the idea of doing it up and then selling it on. She called up a reputable builder, one that had come recommended… a local company whose vans could be seen driving about town, carrying their workforce from project to project.

A representative of the company agreed to come out and give her a price for the work.

Driving from her home to her mother's house, she was short of petrol on the way to the appointment, time slipped away, and she ended up missing her meeting with the builders onsite. But, at the petrol station, she got chatting to a bloke on the forecourt, who just happened to be a builder, or so he said.

Anyway, feeling guilty for missing her appointment, she struck a deal with this stranger to do up her late mother's house. £65K later, her *inheritance* now looks worse than it did before these cowboys ever set foot in the place. They've actually devalued the property while taking her to the cleaners.

Don't buy on the doorstep, or the garage forecourt.

Let's be straight. Even good contractors and good tradespeople can have a bad day, but they will usually come back and try to correct their errors if your envisaged project doesn't turn out as it was intended.

I'm even at the stage where, professionally, I don't want to entertain anyone that is new to me. I've been hurt. I've introduced decent, hardworking people to fly-by-night clients who have turned out to be absolute crooks.

All that glitters is not gold.

Don't buy on the doorstep.

GET THREE QUOTES FOR THE WORK

Let's say that you're a good client with a good track record. You have lots of entrepreneurial stuff going on. In that instance, building companies will be happy to sit on your tender list and bid for any available work. Even though they know a few similar outfits will also be bidding, there's enough of it to go around.

If they win one in three, four, or five of those tenders, they will keep on bidding. You will get your three quotes in (at least).

Happy days.

If you're a prospective client starting from scratch – phoning builders or tradespeople from the phone book or companies you've found on the internet to carry out work on your behalf – then you'll be doing well if you even get one quote back. You'll be doing very well indeed if you get two quotes back.

If you're able to get three back, I'd give you a job myself. As a salesperson, at least. You have the gift of the gab, my friend. You're a charmer.

My advice to all clients – whether companies, homeowners, or wannabe developers – is to be reasonable.

If you can't make yourself understood on the phone or in an email on that initial enquiry, then the quoting builder will likely sense a difficult client (and they might well be right), and you'll get no interest back from them at all.

I know many homeowners are exasperated when builders don't come running when they announce that they need some work done, and I feel the pain of these prospective customers.

I also know lots of builders. I work very closely with them, and even *I'm* often let down by them. It's a shame, even an embarrassment to my profession but, in mitigation, these people are usually busy. If they're hard at work onsite, say cutting bits of timber to size on an industrial saw, then they can't be expected to answer the phone on that first ring. They'd have no fingers left for a start! Accident at work.

So, if people that they know, trust, and love struggle to get a hearing from these busy builders, who are hard at work in a dangerous environment, what chance do the rest of us mere mortals have?

Well, tradesfolk and construction companies need work. There *is* a deal to be done. Somehow.

Be persistent. Be charming. Tell them that you're rich and that you can afford and are even desirous to pay them. Offer them tea and biscuits when they arrive. No one likes a dry job.

Attempt to meet them halfway. You'll pay. They'll do. Treat them like partners, not serfs.

Listen to your prospective contractor. Walk them through your vision for the project, step by step.

Make it worth their while by saying that you're going to move the furniture out of the way, or that you're going to be on holiday, or that you're moving out of the house while the work takes place.

Be a good, intelligent, and reasonable client, and you may just get more than one quote back.

And be realistic in terms of timescale.

The bossier you are, the more excuses you'll hear as to why these people can't come on board.

Bad client = see you later!

Very rarely will you get the contractor that you want, at the price that you hoped for, at a time that suits you, and with the programme duration you expected. Go to the market. Engage and enthuse them, and trust your instincts. Pay a reasonable amount to the people that you want to carry out your work.

If that can be done on the strength of three quotes, that you can sit back with and evaluate, then good on you. If not, remember you've joined a long queue of people who wonder – often in exasperation – how the building industry can continue to operate.

It works both ways.

1. Don't be a difficult customer.
2. Be clear about what you want doing.
3. Assure them that you have the means and the inclination to pay.

That's all that these builders and tradespeople really want to know. And that's how you get your three quotes.

SCOPE OF WORKS

We've all heard the expression 'Mission Creep', and we've also all heard of the expression 'Runaway Train'. That's what can happen when you don't establish the scope of works. Basically, these are the parameters of your project. These are the items of work to be done.

Before we even begin to discuss this, it's worth pointing out that you (or the client or the contractor) should be just as clear about what's NOT included as well as what IS included.

An outline of a simple project might say, 'Build new rear extension to an existing property at such and such an address'.

The scope of works should then list what jobs are actually involved.

Included:

- Demolish existing lean-to conservatory and dispose of debris offsite.
- Construct new concrete floor slab and foundations to building regs approval.
- External masonry walls inc. internal blockwork, insulation, and facing bricks to match existing.
- New plumbing, heating, and electrical connections to new extension area.
- New roof timbers and concrete tile roof coverings to match existing to new extension area.
- New stud partitions, plasterboard, and skim finish plaster to new extension area.

- Two new Velux windows to new extension area above new kitchen/dining area.
- New bi-folding doors and one UPVC external door.
- All as per the architect's drawing and specification.
- NB – we have *not* included for new kitchen fittings or appliances or new sanitaryware. We have not included for any floor finishes or for any decoration works or for any landscaping to the new garden/patio areas.

So, when the client says, 'Where's my new kitchen?', you say, 'Not included. It says so on my quote.'

New sanitary fittings to the new downstairs loo that's been created? Excluded.

Where's my new carpets? Excluded.

Landscaping to the rear of the new patio? Excluded.

Read the bleedin' quote.

Unless properly managed, the tenant/homeowner might say to the foreman or an unfortunate operative onsite 'Where are my new paving slabs? I can't step out of my new patio doors into a pile of mud. You're down to pave it.'

And pave it they might. Unless someone checks the scope of works.

It could and should be there in black and white on the quote letter, clearly stated as part of the list of items of work that have been agreed to be delivered for the contract sum.

Personally, I always exclude floor finishes and decorations when quoting private work. What if the client wants a specific wallpaper from Harvey Nichols, or Shagpile carpet three inches thick?

Exclude it.

In turn, you can always add the installation of the kitchen or bathroom fittings at a later date, or the fitting of a favourite carpet or wallpaper or choice of paint to the walls. Establish the scope of works before you begin. You can add to it or remove items, but you have to have an itinerary that you both understand, with the client and contractor on the same page with an agreed plan of action.

Otherwise, the job, the programme, and the costs involved can run away from all concerned.

THE SPECIFICATION

I largely covered the specification of materials previously, so I'll be brief, but as I'm talking mainly to homeowners in this part of the book, I thought I'd add a few more words here.

On the back of your utterly important scope of works, you need to be *specific* about the standard of product that you want to be used.

Say that you want natural slates on your roof instead of concrete tiles. Say you want 100mm of screeding to your floor instead of 75mm. Say you want Yorkstone paving instead of those ordinary concrete slabs. It all comes at a cost, whether that be an increase or a saving.

And, by all means, utilise your chosen builder's knowledge. You might be groping around (because you might not be an expert) for the right solution to your needs, and they might just happen to know the answer. Have that discussion with them; communication is of paramount importance to overall project success.

Ask for samples if they do throw a suggestion or an idea your way, or ask where this magnificent product can be researched or (even better) viewed. You might easily find some images online.

If you were buying a pair of shoes, you'd choose ones that you liked, could afford, thought suited you, and were more than happy to spend your money on. If you were forced to send someone out to buy them on your behalf, I'm sure you would give them an exact brief of what size, colour, make, and shape to look for. Otherwise, you've got sore feet for the next six months. And no one wants that.

Explain what you want in as much clear detail as you can. Then you might just get what you want. That's your specification.

WHO IS THE PROJECT MANAGER?

We discussed project managers in the professional section. Here's how it applies to you, as the homeowner, in relation to your own project. This could be an extensive refurbishment of your own house or maybe a second or third buy-to-let property that you have in your expanding portfolio.

There are far too many examples, in all areas of life, where people think they know better and get involved in aspects of projects (or maybe even take total control) that should have been left to the experts.

The project manager holds the reins for the duration of the scheme. He or she is the person that people will turn to for answers. They need to give direction, and they need a thorough understanding of everything that is meant to take place. These people orchestrate, motivate, and generate onsite productivity accordingly.

If you, as a homeowner, want to act as PM, are you able to coordinate a potentially disparate bunch of subcontractors? Can you stand up to them if needed? Can you problem-solve and join the dots of your enterprise as it takes shape at each and every stage of its journey?

Can you fight your own corner and ensure that you get the best job your subbie can offer? Do you even want the hassle? Are you going to be an easy touch because you don't really know a poor job from a good job?

If you're not equipped to manage your project, then you're at the mercy of your main contractor or a bunch of subcontractors or tradesmen.

Time to call in your son, aunt, mother, or family friend who can hold these people to account. All I'm saying is someone needs to be the conduit between your aspirations and the end result. That's the job of the project manager.

You may be a natural project manager. You may have excellent common sense and communication skills. You may have a little experience, enjoy the challenge, and not need any external support. If so, you might just be qualified for the task.

If you're unsure, if you're going to flap around and make rash, stupid, or no decision whatsoever when problems arise onsite, then you may not be the best figurehead for your building scheme.

Every project needs a manager. Who's yours?

HOW TO MAKE YOUR FORTUNE IN PROPERTY

Go on. Admit it. This is why you bought the book, isn't it?

Well, as Jeremy Clarkson says on *Who Wants To Be A Millionaire*, this is what I know.

I have always said that I would put my money into property. Unfortunately, I've never had any!

But that's how much I truly believe in it as an investment. Whilst I've not been able to put my own property ideas into practice, I hope that you have a bit more luck.

So, this chapter is very much 'Do as I say, not as I do'.

The first thing to do is to acquire property assets. Do you own your own home? Can you remortgage it? If your house has gone up in value by fifty grand in the last five years, could you extract that value from your lender and use it as a deposit on another house? Maybe a buy-to-let.

Now you're moving up the ladder.

Can you see value in bricks and mortar? Do you trust your own judgement? If not, have you got the ear of an expert, friend, relative, or an independent agent who can point you in the right direction?

Actually, beware of either the friend and the relative version if you can. If it all goes pear-shaped, it can make for awkward afternoons and evenings at those birthday and Christmas gatherings.

Gut instinct. That's what you want. An educated guess, at least. Research helps. Is there demand for what you are planning to build, buy, and possibly rent out? What are the selling points of your potential purchase? Even better, what are the *unique* selling points?

Is the place you have your eye on better, or at least as good as the surrounding properties?

What about transport links? Good schools nearby? Parks and other amenities? Basically, why would anyone want to live there?

Can you shout it from the rooftops, the product you're ultimately going to take to market? Can you extol its many virtues, or is it really a dead rubber that you're trying to shift onto someone else?

I once knew a girl who was trying to sell her place in London. It was a ground floor flat in a high-crime area. She'd been burgled five times in the previous year. She'd had enough. Wanted rid. So she put it up for sale. A prospective buyer asked her if there was much crime in the area.

'Nah,' she replied.

'Ever been burgled?'

'Nah,' she said again.

And the unsuspecting victim took the property off her hands.

Has the purchaser been burgled since? Quite possibly. Has it gone up in value since she bought it? Quite probably. And when the next prospective purchaser asks the owner about crime in the area, what do you think she is going to say?

Let the buyer beware. Caveat emptor.

In the *Daily Mail's* 'Me and My Money' column recently, a singer with the pop band *Liberty X* was the focus of the interview.

'Do you own any property?' she was asked.

She said she did. She'd purchased an apartment in Leeds for £210K in 2006. The current value? Exactly what she initially paid! No movement in value whatsoever in over a decade and a half.

In the meantime, the house I live in has more than doubled in value over the same period, as have all the best-bought properties. On average, you could and should have doubled your money on your home in that time.

What was the problem with the place that the former singer bought? Too much of the *same thing.*

There was no cultural or spiritual value to the enterprise, but people at the height of that particular property and economic boom were suffering major FOMO. The fear of missing out.

Buy when everyone is selling. Sell when everyone is buying. Take a long-term view of things.

Your home can be an investment, but what we often forget is that it's your home first and foremost, the place where you rest your head each night. You need to enjoy it. Everything else – including monetary uplifts – is an added bonus.

Ask yourself how much time you want to devote to your property portfolio. You may buy your own house and stretch to one or two others. Keep it manageable. You can get on with your day job then.

If you are actually looking at having ten or 20 properties under your umbrella, then that is a full-time job. It will have its own challenges, which means that it will involve a fair bit of stress.

Still, if you know what you're doing, and this is something that you really want to do (and you have the necessary skills), then it will mean you are your own boss and can come and go as you please.

As long as the rental returns cover your mortgage costs and leave you with a salary at the end of it, which in turn covers your outgoings and pays for your lifestyle, then all's well and good.

I know a lad from my 5-a-side football days. He makes a good living from property. Here's how.

His dad had blazed the trail. He'd bought, refurbished, and let out properties to students in a part of a city close to the main university campus. The area also had a number of cool bars which attracted the students, so they were easily let. Anyway, when this lad reached a reasonable age, say 20 or 21, his dad gifted him his first property outright. It might have cost the old man fifty grand, and his kid was now on his own.

With equity from the first house, he could now remortgage or use the rent to fund another property.

Onwards and upwards.

He probably bought one or two terraced houses a year. Did them up and rented them out.

He told me that he did them up to a good standard as tenants seemed to respect the place more.

So, let's say that this lad buys just one property a year from the ages of 20 through 50. By the time he has reached his half-century, he owns 30 properties.

Say the first of those dwellings cost £50K. The mortgage has been well paid 30 years into the future. That house is now worth maybe £200K.

He can retire at 50, and hand the management over to an agency, of which there are plenty willing to take on the role.

He can then sell one house a year, slowly divesting himself of his portfolio, and if he lives to be 80 years old, he'll have £200K a year to live on, with those figures rising annually as long as he's bought well and the properties keep rising in value.

Unless you've bought in an area that then degenerates into an urban ghetto, most houses should steadily improve on their worth.

As a surveyor, I meet all sorts of people onsite. I once met a subcontractor on a job and we got talking. He was installing suspended ceilings on one of my projects. He said that being a boss was not without its pressures. He had to find £5K a week to cover the wages of his operatives. A mate in a similar situation told me he'd once had to pawn his and his wife's jewellery to pay their staff on a Friday at a time when they were still building their business and waiting for client funds to land. Being a boss, I guess, is not for the faint-hearted.

Anyway, this subbie said that, wage pressures aside, his business occasionally made good money, and it had allowed him to purchase about five different properties. His modus operandi was this…

Buy in the very best areas. These would be larger, detached houses that attracted (and could only really be afforded by) the middle class or the well-off. Then, he would only ever rent them out.

Now, I know that most of us consider rent to be dead money, I don't disagree for a second, but the fact of the matter is that there will always be a rental market. Some people can't afford to get on the property ladder and have to rent against their better judgment, while some people are transient and can write off the rent as the cost of passing through. It may be people working on a short-term contract or whatever. There has always been, and will always be, a vibrant rental market.

So, this working construction guy buys houses that tick lots of boxes, in nice areas, and rents them out. If he can do that year in and year out, at some point the mortgage will be paid, and he'll own the building outright. He got it for free! Or rather, for the cost of keeping an eye on it and maintaining it, but it's still a small price to pay for a house that will probably be worth upwards of half a million pounds by the time he reaches retirement age.

Have something that you can sell again and again, and yet you still own it. That is the path to success.

The best advice I've ever heard is not related to construction or property (I actually read it in the biography of a film producer), but I think it applies to every avenue of your life, so it is still relevant to your burgeoning property dream if that is the way you want to go.

So here goes.

When have you ever done something well? When have you achieved a notable success? Hopefully, you'll have had that experience at least once in your life, otherwise this advice is purely academic.

Okay, so you *have* had some success. Well done. Now, this could have been in any field. Was it an exam that you passed? Your driving test? A job interview?

The point is, you did it. You succeeded. How? Why?

There was a reason.

Go back to that place. What did you do? How did you get there? Was it preparation? Was it by not really caring about the outcome and therefore you were more free-spirited, showed more of your character and personality, and that's what made you stand out from the crowd? Was it simply hard work? You put the effort in; you burnt the midnight oil. What were the steps that you undertook that then resulted in that successful outcome?

Find the reason. Write it down. Then repeat. You can do it again. After all, it worked before.

The same is true for your property empire. Which ones did better than others? Which ones turned out to be a nightmare or not worth the effort? Examine the good ones. Learn. Repeat.

Let me tell you about my cousin.

He's an interesting dude and long had a hankering to work in property. Formerly in the army, he'd even worn the famous bearskin as part of the Queen's Household Cavalry. After leaving the military and working in security, he still had that property dream.

Finally, he made a start on it. After all, we all have to start somewhere. Twenty years later, he's still at it. What he doesn't know about property right now isn't worth knowing.

He has many strings to his bow, his fingers in a lot of pies, and he can offer a range of services across the construction and the property industry to his ever-growing list of clients.

He's trustworthy. After all, he was given the job of guarding our former Monarch. That says it all.

He offers good advice. *Making other people rich*, he calls it. He does get a small percentage of it all, though, which makes him a decent living as he has a full order book of different projects.

He once had a neighbour in the picture-postcard village where he lives. They were looking to move on and had been trying to sell their property for absolutely ages. Lots of viewings, but no offers.

My cousin organised a minor refurb. A spruce-up with a bit of paint and new carpets. Overall, the work cost the couple five grand, and his own markup was ten percent. Too cheap in my book, but he's playing the long game. (He gets good word of mouth and has become known for being more than reasonable.)

The house sold in an instant as soon as the work was complete.

Some people, when it comes to viewing a property – their potential new home – have little-to-no imagination. By sprucing the place up, my cousin had effectively taken the blinkers off for them. He'd introduced light where there was previously only dark, and did the thinking for the prospective buyers who would now step through the door and see the real potential that the place had to offer.

The existing homeowners were gobsmacked. Thrilled, in fact, to have finally seen their property sold. They could now get on with the rest of their lives.

Following on from that success, my cousin then reached out to friends and family. He said that if any of them had any money to invest, he would go and find suitable properties and put their slush fund to good use by pointing their savings in the right direction. In bricks and mortar.

One bloke he knew, an old schoolmate, was interested in the idea but also a little cautious. He asked if they could go halves on the first investment that my cousin proposed to him. After all, if it was *that* good a thing, why would my cousin not also invest?

My cousin said yes. That removed all of the insecurity that the prospective investor had about the scheme.

And they both backed a winner. It's called putting your money where your mouth is.

I have a friend called Sue. She works as an estate agent and has bought a few properties herself. She buys places with rental potential. That's all she does. She doesn't buy and then instantly sell, which is called flipping. You'd probably need to do that three or four times a year to keep you in that property developer's salary bracket of at least £100k to flaunt down the local wine bar.

Sue's view is that it is far, far better to take a long-term view.

Initial cost of borrowing. Expected rental returns. Manageable number of properties. If you have ten properties that each earn you £500 more than the mortgage outlay each month, then that's the same return as the 'flipping' model above, except you still own the damned things.

If you've bought well, then a property's worth will also rise each year. In ten years' time, you might have paid the mortgage off, and yet you still own it. Now that's a time to sell if you want to.

And who does my friend Sue let to? Footballers. Wealthy individuals. 'I've paid a certain Premiership football manager's rent for the last six months!' she once said to me. How many of us can say that? (She did actually name him, but I'll spare everyone involved and not repeat it here.)

Not that the gentleman was unable or unwilling to pay, by the way. Simply, his employers, a famous football club who were covering his living expenses, were always tardy when it came to accommodation fees. They always paid in the end, and Sue could afford to wait.

So, bricks and mortar in good areas. You can't really go wrong.

Over the last half a century, while average salaries have risen 60-fold, average house prices have risen 180-fold. Whereas house prices were roughly worth three times the average annual salary 50 years ago, they are now about nine times more expensive.

Your home, and real estate in general, has basically outperformed the basic wage by 300 percent over a sustained period of time.

I've always said I'd put my money into property. Now, where can I get some money…

12. COMMON CONSTRUCTION MISTAKES

WHY COSTS OVERRUN

I'm a cost surveyor, not an accountant. As Albert Einstein once said, 'Not everything that counts can be counted, and not everything that can be counted counts.'

If I've mentioned that already, well, as the old saying goes, 'If something is worth saying, then it's worth saying a thousand times. And anyway, who's counting!'

Why do building costs overrun?

Poor preparation. Poor delivery. Unrealistic expectations. Ignoring reality. That's the size of it.

I once took a phone call from a random bloke who had called our construction company wanting a bit of advice. Reception put him through to me, and the young man said that he was thinking about buying a particular property. The house that he had his eye on needed quite a lot of work. He wanted to know how much it might cost to refurbish it. This was around 2015, so the costs noted below have increased since then. But, anyway…

I'd never even seen the place, but I knew a bit about the city in which we both lived, and I got the message that this was probably going to be his first step on the property ladder. This was going to be a doer-upper, and it wasn't going to cost him the earth to purchase the building in the first instance.

'Would ten grand cover the construction work?' he asked. He was talking about a general refurbishment; a bit of a spruce-up to a small terraced house, which was standard fare for our company.

'I reckon more like twenty,' I replied.

'Not ten?'

'Look mate,' I said, 'you can spend just 50 quid on it if you like… but the place will still look crap.'

Let's do some maths. The bloke is probably hoping that a team of builders will be in there for about six weeks, busting their guts to improve the place on his behalf. He probably wants a small army, but let's just say we have four men working away for four weeks and see what that gives us.

Four men, consisting of two experienced tradesmen and two young lads doing the donkey work, are probably going to need to take home about £3,000 a week between them. A grand each for the big guys, and half of that for the lads doing the labouring. For four weeks, that's £12K.

Then there's the materials, new kitchen and bathroom fittings, new joinery, plasterboard, glazing, internal and external doors, plus the paint, etc. And not forgetting the skips and the scaffolding. And, if you're not project managing it yourself, there will be the building company's overheads at about a 30% mark-up.

This bloke wanted it for ten. I told him at least £20K. And that was only if he got really lucky.

So, unrealistic expectations at the outset! There's the first potential (and common!) reason for a cost overrun.

Then there's mismanagement. A project that idles interminably or possibly runs away from you.

Tradespeople will happily sit onsite and do very little work as long as they are getting paid for it. Therefore, you need to manage, motivate, direct, and pay appropriately for what you are getting. No blank cheques. No 'what are we doing today, boys?' Direction. Focus. Motivation. Get value for your money.

And if you are a contractor, *give* value for money by offering a subbie a set price to do a certain thing. Be clear what that is. Put them on a price to deliver it. No one ever wastes time when it's *their* clock that's running.

With costs rising across the board in the construction industry in the present economic climate, whether that be as a result of Brexit, continuing ripples from the Covid-19 lockdown, or the cost of living crisis, these are all just easy excuses when it comes to discussing the finances of any construction project.

It's a competitive market, but if you're a contractor and you caveat for every possible eventuality, you'll price yourself out of every job. Embrace some risk. Cost for it. Ignorance is not bliss when it comes to pricing work.

By all means cover your arse and even tell the client that's what you've done. Then, if you win the contract, you have an understanding partner who expects you to deliver because they know you've analysed, prepared, and costed for these likely events, and you think you can still meet their budget.

They can choose the more attractive lower offer if they like, but you know how that story is likely to end. Badly. The costs of their project will overrun, because they chose to ignore reality.

Also, if your project is stagnating due to a lack of leadership or a funding shortfall, then this will adversely affect *your* costs. It might look like nothing is happening onsite, and that might well be the case, but there will still be the cost of borrowing any initial funding, plus any plant or equipment, cabins and the like that are on weekly hire onsite, plus any site management or behind-the-scenes staffing costs. Your overheads will still be mounting up even while your build scheme is on hold, and this is true whether you are the client or the contractor.

So, even though you don't appear to be spending, your expenditure will still be going up. Better to plan and budget wisely from the outset, and then push on with all necessary speed to completion.

Costs also overrun because of a lack of forethought at the estimating stage. Have you included for absolutely everything that your project needs? Plug those gaps; build the thing in your mind's eye before you settle on a price. Include a contingency if possible. Don't catch a cold on your budget.

If the construction phase of your project is going to last for several months or years, then unless you can procure all of your materials at the initial cost, and you have adequate storage space for these items, then you are going to have to pay the inflationary increases that will occur along the way. Have you allowed for this? If not, then there's a cost overrun right there, unless you allow for those increases within the terms of the contract, which is permissible and advisable.

Another reason for projects coming in over budget is purely for political reasons. That's politics with a small 'p'.

Who gave you those costs in the first place? They might have given you a good-luck story simply to protect their own jobs. Ask yourself who prepared the budget estimate, and why. Were they just trying to please their client (which could be the homeowner or even the supposedly streetwise contractor)?

Do a robust analysis of your budget at the *outset*. If you don't, then you shouldn't be too surprised if the end cost looks different to the one you envisaged at project commencement.

WHY PROJECTS FAIL

I recently had the task of conducting an independent review of a failing project for a construction company. It had become apparent that the firm was in danger of losing a considerable sum of money on the scheme after almost three years of supposed endeavour.

After casting a fresh pair of eyes over the project, the news that I had to give them was all bad. It was a failure on every level. If there are ten criteria by which you might measure the success and failure of a particular scheme, then there would have been ten red 'X's and no green ticks across the various items.

Underpriced at the very outset, there were just so many different elements of the build that the estimator had simply ignored, failed to consider, and ultimately did not include. To me, it looked like a 'back of a fag packet quote', and it failed to take many of the items of work into consideration.

Although the project was underpriced, good procurement could have recovered some ground. Perhaps the team could have bought the right stuff at a good price? Sadly not. It was the wrong stuff at over-inflated odds in many cases!

More red 'X's appeared.

Awful management and supervision of the scheme; poor quality of labour; no commercial management (i.e., securing funds from the client as and when they became due).

After almost three years onsite, at circa 90% completion by this time, the contractor's QS had only claimed about a third of the *preliminary costs.* This is your management, welfare, cabins, skips, scaffold, plant and equipment, etc. It should run in tandem with the actual building programme. So, if you're nine-tenths of the way through, you are entitled to nine-tenths of the cost set aside for this element of the work. Yet they hadn't claimed it. Bizarre. Outrageous, in fact.

How about the quality of the work?

Dreadful. It was almost at the point where the client could (and they actually still might) condemn it and refuse to accept it. There were damaged installations – left, right and centre – and while the product, an industrial insulated roofing panel, could have been said to serve a purpose (an excuse that the main contractor was desperately clinging to), in my view it was unacceptable to expect a client to accept something 'new' that was clearly crumpled and dented, bent out of shape in fact, in so many places.

For example, a new car might still drive and therefore be deemed 'fit for purpose', but you want your new car, home, or anything else for that matter to look the part when you first take possession.

Health and Safety for this scheme? Another red X. Scaffold not checked and not tagged, therefore illegal. Ladders that weren't adequate and should have been cordoned off, but which were still active. I was told that the Health and Safety team had only been to the site twice in three years, and not once had they got up on the roof, which was the main area of work. Just shockingly bad. Piss poor.

Projects fail because they get many of the essential criteria wrong. Initial estimate, contract terms, procurement, delivery, cost control, site supervision and project management, health and safety, workmanship and quality, client satisfaction, and profitability.

When I came to report the results of my independent review, I stated quite clearly that the project was without leadership. I said that the huge predicted shortfall in revenue was actually a best-case scenario at that stage. The ship was listless and could well end up on the rocks. Unless someone took hold of it and steered it to completion – in as short a time as possible – the costs to complete would continue to rise.

As former White House Chief of Staff Erskine Bowles once said, "Leadership is the key to 99 percent of all successful efforts." Take that list of ten criteria. Take them all seriously and do them all well.

Think of yourself as a Premiership football manager. Get ten green ticks, you're winning the league. Get nine wins and a draw, you're still doing extremely well. But get more red 'x's than green ticks and you're either mid-table or heading for relegation.

That's why projects fail.

POOR DESIGN / POOR QUALITY

I was recently sat at a window in an upper floor of the famous Shard building (yes, back there again!), admiring the London skyline, including the historic Tower of London, Tower Bridge, the 'Walkie Talkie' office block, as well as others soon to join the skyline judging by the proliferation of tall cranes across the capital.

A couple of old buildings were in the process of being dressed in scaffolding, about to undergo some major restoration work. So many of their ilk had previously been demolished to be replaced by newer modern constructs, and I could only hope these original buildings – about to get a makeover – would somehow retain some of their impressive features once the renovation work was complete.

Back in The Shard, ahead of a design meeting, there was a bit of coffee table reading in front of me. Inside a magazine, I came across an article written by an ethical travel company that had made their business out of hiking, climbing, and making and selling outdoor clothing.

Along the way, the company had opened up a series of retail outlets and were now leaving their footprint on the built environment. What was interesting was the rules they adopted as they kicked off their muddy boots and stepped into the world of bricks and mortar; they advocated for the optimisation of aesthetics, function, and responsibility.

Along this line of thinking, this ecofriendly company identified various rules of thumb.

Never build a new building unless it's absolutely necessary. Instead, buy used buildings and do them up, re-using old construction materials and recycled furniture wherever possible.

Try to save old and historic buildings. Any structural changes that you make should honour the historic integrity of the original building.

If you have to build a new building, make sure that its aesthetic life expectancy is at least as long as the expected lifespan of the fabric of the building.

Anything that is built should be repairable and easily maintained.

Buildings should be constructed to last as long as possible, even if this means they cost a little more to build initially.

Your building should also be unique and reflect the history and the natural features of the area. These local values, identity, and attributes should be reflected and honoured in the building that you are about to add to the existing environment.

In other words, what you add should also improve things. Otherwise, you and your building are a detriment to the local community.

Design well; design brightly; build quality; quality lasts. Do these things together, and then reap the rewards for many a year to come.

POOR COMMUNICATION / DIFFERENT GOALS

As soon as a project goes live, the clock is ticking, and the meter is running. This is where good communication can make the difference between a successful outcome or an abject failure.

I don't believe I've ever had such a thing as a wasted site visit. The stuff I've picked up in so-called general conversation with contract managers, subcontractors, and the operatives carrying out the work has deeply informed my knowledge of schemes that my employer or clients are engaged with. It's good to talk, and to listen.

Someone mentions that they are on holiday next week. You remember that you've got an expensive crane booked in for next

Wednesday to facilitate their work, but they're not going to be there. That crane needs cancelling. Fast!

If you don't know about it, you can't act.

It is expected of professional and conscientious people that they communicate effectively. It is the oxygen of any operation. If someone chooses to ignore any advice and information that they are given, then the blame (or the subsequent demand for answers) for any calamity that results should be laid squarely at their door. People burying their heads in the sand and not passing on important information is essentially a dereliction of duty. And the outcome can be detrimental to your project's success.

In a nutshell, according to a 2023 online report conducted by Pro Crew Schedule, entitled *The effects of bad communication in construction*, only 37% of projects with minimally effective communication finished on time. Projects with highly effective communication saw 71% come in on time. That's almost twice the success rate.

Now, some people, certainly down the food chain, act on a need-to-know basis. These guys need guidance. They need to be told, in no uncertain terms, what it is that they are there to do. There's no harm in that, as they aren't going to put their heads above the parapet and make decisions. Meanwhile, those privy to the goals of the enterprise should ensure that news filters down. Square pegs for square holes from thereon in, and all delivered through the essential medium of effective and timely communication.

That's why we have meetings. Pre-start meetings, progress meetings, opportunities for everyone to get on the same page. Get-togethers to discuss the issues of the day, the path to completion, the obstacles in our way.

The same principle applies to any small domestic project. Talk to the contractor, the boss, the lads onsite. Does everyone know that they are meant to be doing such and such a thing but *not* touching that bit over there?

Talk until you have the reassurance that everyone knows what they are meant to be doing. Obviously, at some point, the physical work needs to take precedence, so don't talk the job to death, but once those involved have their roadmap, let them get on with it.

Communicate. Motivate. Ensure everyone has the same goal.

13. RECENT CONSTRUCTION DISASTERS

GRENFELL TOWER

Grenfell Tower was erected in the affluent borough of Kensington and Chelsea in London in 1974. Part of a nationwide craze for high-rise living at the time, the block would deliver 120 apartments on a single site. Despite the desirable postcode, the immediate area around the plot was rife with issues, including anti-social behaviour and vandalism. The former terraced houses that gave way to the tower were considered a veritable slum, not in keeping with the surrounding properties. After their demolition, the Grenfell block was raised in their place.

The development had a troubled early existence, and in a poll conducted by the local council in 2012, over 80% of the residents complained about the state of the living conditions at the tower. The council agreed to a programme of refurbishment, scheduled to take place between 2015 and 2016, and the main items of work were going to be the gas central heating supply, the windows in each of the apartments, plus improvements to the thermal performance and the overall aesthetic appearance of the entire block, for which the proposed solution was to wrap the building in a new curtain-wall facade.

All well and good.

The parties involved in the planning of the scheme included the preferred building contractor, the local council, and their professional consultants (architects, building surveyors, and the like). The initial proposal for the external cladding was for an aluminium product containing a robust, fire-resistant core, but the cost for the scheme – including the upgrade to the windows and the central heating – was more than the client, which was ultimately the local council, could afford. It was time to reduce either the scope of works or the specification of the materials to produce the required cost-saving.

Some bright spark on the advisory panel suggested using a cheaper curtain-walling system. The same aluminium façade, only the inside material would now be more combustible – not the more fire-resistant one that was initially specified.

Did building regulations prevent this? Sort of. But only if you looked closely and interpreted the rules correctly.

Did the manufacturers and suppliers of the proposed product stipulate the exact criteria for use? Not really. After all, you want to sell as much of the stuff you're making as possible, so why put off potential purchasers? They sold it anyway with few, if any, questions asked.

What they should have done was put a huge disclaimer on the datasheet for the insulation. A datasheet basically says what the material is, what its uses are, where and when it is appropriate to use it, and more. Such information is generally available and supplied with almost every building material.

'Not suitable for use on buildings over ten metres tall.' That would have been a very appropriate heading. In fact, Arconic, the company that sold the cladding sheets, does issue a 'diagram' that illustrates how PE panels are only suitable for use in buildings up to ten metres tall. But then, included in discussions with the project management company (an outfit called Harley) and the principal contractor (Rydon), they were asked to provide samples of various PE materials for the Grenfell Tower project. Was the clue not in the word 'tower'? Could the supplier possibly have said at that point, 'This won't work for your scheme as it's a flammable product not suitable for tall buildings,' etc.

And why ten metres?

That's the maximum height at which a fire engine can effectively tackle a blaze from the ground. Their pumps and hoses can't really propel their flame-quenching streams of water any higher. Not rocket science, is it?

Many people have argued that datasheet information, available for most construction products, should carry legal weight. In the example above, a disclaimer could boldly state that 'It is illegal to install this product on any building over ten metres tall.' That would grab people's attention. Let an architect or a contractor try to install it then.

Still, the law doesn't require such absolutes, so we are left with a situation whereby the contractor wants to win the job and offers the client a cheaper alternative. The client, meanwhile, thinks they are working within the accepted rules and spending their limited budget wisely, as well as looking after the interests of their council taxpaying customers. At the same time, the manufacturer is merely making a sale. As far as they are concerned, it's the responsibility of the end-user to make sure they use the product wisely.

Do you get my drift? There are lots of gaps here into which everyone can fall.

What should have happened, quite clearly, is that someone should have spoken up. And yet no one did!

So, the refurbishment took place. New windows and an upgrade to the gas pipes and central heating. The building was then wrapped in an aesthetically pleasing blanket. A flammable one. And all within the available budget. And just about within the rules.

Looking back, it seems absurd that no one said, 'What the hell are you doing? This is dangerous,' as it was being installed. No architect, no engineer, surveyor, building inspector, contractor, or committee member for the local council raised a single concern at this juncture as far as I'm aware.

These planning meetings, pre-start meetings (and I've been to hundreds) usually follow an itemised agenda, and there are also open forums in which the so-called experts and stakeholders can discuss everything in the round. Yet no one saw the danger. No one raised any objection.

At 54 minutes past midnight, on the 14th of June 2017, a resident in Flat 16 on the 4th floor of the Grenfell Tower made a 999 call to report a fire in his flat. The cause of it? An electrical fault in a large fridge-freezer. Just five minutes later, the first firefighters arrived on the scene to tackle what they believed to be a self-contained blaze in a single apartment.

But with the fire brigade on-scene, and only ten minutes after they arrived, the fire reached the exterior cladding. Maybe the heat from within the flat cracked a window, or some cavity allowed the flames to migrate, but the fire exited Flat 16 and then set alight the internal combustible element – the core layer of the aluminium facade. The PE-type insulation that formed the inner skin, the thermal-improvement element of the outer panel, was completely flammable. And now it was on fire.

And from that point on, all hell broke loose.

Fire, like water, acts according to its own laws. Once aflame within the foam insulation, which was secreted between the original external face of the tower block, presumably a pre-fabricated concrete panel, and the new all-gleaming aluminium facade, well the fire then went wherever it wanted, or where nature and the prevailing winds directed it.

Up, around, over, and down. Left, right, here, there, and everywhere.

And it also went back in. Back into the apartments. Because as it coursed up the exterior walls of the building, the heat cracked windows, and the flames licked at the curtains, then billowing in the blustering wind, and sneaked, sought, and grasped whatever inlet it could find. No opportunity was missed.

As the people in all the other apartments called the fire service to report smoke and fire and asked what they should do, the initial advice was for them to stay exactly where they were.

Each residence had fireproof doors, and there was an expectancy that they would provide fire resistance for long enough for the people inside to be reached by the fire service. Why risk venturing out into the communal areas when you could just sit tight and let the professionals arrive and put the encroaching flames out?

This standard operational advice, dispensed in good faith by members of the 999 answering service, was only amended much later on in the night and the wee small hours of the morning as the extent of the unfolding tragedy began to be grasped by all involved.

The people who fared the worst were those on the upper floors. By the time the fire service advice was changed to, basically, 'run for your lives', the building was a blazing inferno.

One resourceful resident of the tower block did do something drastic that saved their life. They flooded their own apartment.

Realising that they could not leave, and that they couldn't just do nothing, and that the building was probably doomed anyway, so a little more damage wouldn't matter one iota, they turned on the taps in the bath, sink, kitchen sink, etc. Then, they put the plugs in the holes and allowed them all to overflow.

Wet floors. Wet beams. Wet carpets, Wet everything.

And when the flames reached that upper-floor apartment, guess what? Those flames moved up and around it because there was nothing for the fire to take hold of in that thoroughly soaked environment. They thought quickly, took drastic action, and lived to fight another day. Sadly, so many did not.

Over 70 souls were lost in that tower block on that dreadful night.

It took just 18 minutes for the devastating fire to spread from that fourth-floor catalyst to the roof. The polyethylene cladding core provided fuel for the flames; the 'aesthetic' cladding itself was described as 'combustible' in the report issued for Phase 1 of the Public Enquiry into the disaster.

Allow me to make this abundantly clear.

Don't wrap buildings in flammable blankets.

If you are any sort of construction professional, speak up in the design meeting or at the pre-start meeting or progress meeting when issues regarding health and safety, and their blatant disregard, come to the fore. Just speak up!

As for the regulations, don't bury something as important as this issue within requirement B4 (1) of Schedule 1 of the Building Regulations 2010. Who on earth is going to notice it there?

It always seems to take a disaster to shine a light on fundamental flaws in any process. How many could be prevented with the right rules and regulations in place, and a bit of common sense by industry practitioners when they come to design, consider, and then implement the act of construction?

The manufacturer of the cladding material that facilitated the spread of fire at Grenfell has now discontinued that particular product, citing the cause of their decision as the confusion around fire regulations in different parts of the world.

A nice line, that one. Draws a line under the issue. With hindsight, I'm sure they could have been clearer about where and when their product should have been used. And others should have done their due diligence as they came to consider its use. But, as always, there are just too many people looking after their own interests.

And there should be a law against that.

BUILDING ON FLOODPLAINS

There are currently over 5 million homes in the UK that are built on land that is liable to flood. That figure is set to double in the next 50 years.

Around 20,000 of the new homes constructed each year are built on land that has a high flood-risk rating. Does that make any sense to you? It certainly doesn't to me.

According to a survey by the Environment Department (DEFRA) in 2024, over half of all local planning authorities rarely or never check whether a new development complies with flood risk measures. And the misery goes on.

What is a house? What is a home? What benefits does your abode provide for you? I would say shelter and security. But how could you possibly relax when every evening's weather bulletin has the potential to ruin your day? How do you sleep at night when an extreme weather event could literally come flooding into your home?

I know humans seem to tolerate a bit of risk in their choice of location – why would anyone choose to live in San Francisco, for example, when it sits on the San Andreas fault line and could be subject to a catastrophic earthquake at any given moment – and yet people do.

But, come on. Would you buy a house on a floodplain? Would you buy a house that comes with the highest category of flood-risk rating? I certainly wouldn't. And nor should you.

So, who buys them? And why do we build them?

Well, when the government of the day makes promises to provide hundreds of thousands of new homes every year to ease the housing crisis, and that responsibility filters down to local councils to deliver the homes to meet the demand for these properties, that's the underlying reason right there. Demand and political pressure.

The only government directive regarding building homes on floodplains is that it shouldn't be done unless it's unavoidable. There's the get-out. Who's to say what is and isn't unavoidable? A simple email to say that no alternative sites could be found, and certainly none that could be purchased within the available budget, should be enough to satisfy that 'unavoidable' condition.

Basically, the shortage of housing outweighs any reservations on the ultimate suitability that might stall a proposed development. We take risks because we simply have too many people in the UK who need accommodation.

I suppose, if you want to look at it in base terms, the absence of housing means, essentially, homelessness. So, proportionally, a home at a higher risk of flooding is better than no home at all.

But I'm being ridiculous here.

Surely, it is possible to build houses in areas that are unlikely to flood, even in extreme wet weather conditions? I think it is incumbent on developers to explore every avenue, and indeed every vacant field, in less flood-prone areas before any local council or planning committee green lights a development that comes with such an obvious basic flaw.

After all, flooding is absolutely devastating for those affected, and people can spend a year putting their houses back in order after such an event, only for the same thing to happen all over again. It's heartbreaking. Why bother?

And this happens in towns and cities that are well-established, and which have been a dwelling place for human beings for hundreds if not thousands of years.

So why them? Why now?

Well, we are currently experiencing a period of climate change where the planet is warming up, sea levels are rising, and extreme weather events are occurring with greater frequency. Regardless of your thoughts on the *cause* of all this, climate change is a stark reality. It's a fact. It cannot, and should not, be ignored.

It may be that the planet is merely going through one of its cycles, and this may all calm down naturally and even reverse in a couple of hundred or a thousand years' time. I don't know! It doesn't really matter anyway because none of us reading this book will be around to see it. All we can do is deal with the here and now and the foreseeable future.

And in the here and now, we are short of purchasable land to build the number of homes that the country needs. Hence, we flirt with danger and throw common sense to the wind and build houses on floodplains. And when those houses are built, people queue up to buy them. Just don't expect to find me in that queue.

If we are going to ignore the blindingly obvious (i.e., don't build homes on floodplains), then what steps can be taken to mitigate the risks?

Well, if we are building in coastal areas, then sea wall defences are an option. I've been part of a project where huge concrete boulders were fabricated and dropped into the ocean to form an unnatural barricade.

Still, I can't forget the Japanese tsunami event that destroyed a nuclear power facility. The designers of the power plant estimated that an underground earthquake might generate a wave of six or seven metres, so they built a protective wall eight metres high. Well, forgive me, but that's only a small margin for error. Anyway, the earthquake happened, the wave was higher than the wall, the result was devastation, and I bet they all thought 'should have built it higher'. I'm sure that if I thought that a seven-metre wave might be coming, I'd build a wall 15 or 20 metres tall!

We *can* build suitable defences.

We can also look at substantial drainage systems. You know that the water is coming, so plan in advance to gather, direct, and then contain it in tanks, fields, or underground systems away from the actual housing or the streets that serve and connect them. By the way, all of those tarmacked highways and concrete pathways aren't porous. The water that lands on them will run and run until it eventually finds a home. Possibly yours.

Predict, manage, and divert.

I saw a news report once about a new homebuyer whose garden was unusable. It was built so close to a flooded plain and in a high water table area that his back garden was basically a swimming pool. Well, that's not entirely true. A swimming pool would be a desirable feature for most of us!

The garden was a sodden quagmire. A boggy field. The newly laid turf was a squelch-fest. There was no chance of having a kick-around out back with the kids. You couldn't even let the dog run around in it because, once invited inside again, the family pet would be leaving muddy paw prints all over the furniture and flooring. So, basically, the new homeowner didn't have a garden. Not one that he or his family could use, anyway. And that was a brand-new home. Knowingly built on a floodplain. Caveat Emptor; let the buyer beware.

River diversions can bring solace to some towns, but once those tributaries emerge and re-engage further downstream, they have the effect of swelling the deluge in that location, so while the (usually) upmarket town (upriver) is less likely to endure flooding, the less affluent town bordering the same river route farther down the same waterway is now more exposed. And occasionally gets flooded.

Don't build on floodplains. Don't buy on floodplains.

Now, here's a funny story. Again, watching the news. A scene of devastating flooding where a whole newbuild development at the edge of a village was almost washed away.

The TV news had a reporter on the spot who interviewed one of the poor, unfortunate residents. The man had watched half of his furniture sail away down the street. He was totally gutted.

He'd been in the house a couple of years. Didn't think that something like this would occur.

'I never thought it'd happen here,' he wailed into the camera.

The reporter finished his piece by saying, 'This has been John Smith for the BBC, live from West Marsh!'

West Marsh!

Don't buy a house in a place called 'West Marsh' and then complain that it's flooded. The clue is in the name. It's just common sense. But I guess you can't buy that either.

THE LEASEHOLDER RIP-OFF

Congratulations. You've just bought a house.

Or did you?

Some purchasers discover, after the deal is done, that they don't own the land their building – their house, their home – actually sits on. They are leaseholders. This means the lessee has the right to occupy the property for a stipulated number of years. They're paying twice: once for the lease and once for the ground beneath it.

As such, they will forever be at risk that the ground beneath their feet could be sold if the leaseholder still owns the plot that their brand-new home is built on. They might decide to double, triple, or quadruple the ground rent, or sell the leasehold on to an outside party, who may decide to double those rates again.

Historically, there have always been houses that were sold as leaseholds. Rich landowners allowed people to build homes, but maintained ownership of the land on which they were built. Feudal instincts abounded. In the 21st century, you would think that such practices and such ways of thinking would be on the decline.

No.

Some engaged individual – an executive or a marketeer at one of the major housebuilders – no doubt saw the opportunity to revive, expand, and exploit this ancient practice and impose it on unsuspecting housebuyers in more and more numbers. Smelling blood and opportunity, their competitors were quick to join in the frenzy. Vultures, a great many of them are.

By 2015, almost half of the newbuild houses being sold in the UK were leasehold instead of freehold. And these important details were often buried in the small print of the legal pack.

The salespeople and the sales literature being touted by these often household-name housebuilders, certainly weren't shouting this tiny yet crucial detail from the rooftops. Instead of buying and owning your new home, unsuspecting purchasers were still essentially tenants.

Homes were being sold as 'virtually freehold'. Which means what? Not Freehold. That's what. You don't own the damn thing.

Anyone spotting the small print and probing the possibility of buying the freehold was told that it could be purchased in a couple of years' time for a couple of grand. They were on a 999-year lease anyway. You'll be dead by the time you're a hundred years old. What have you got to worry about?

Maybe the fact that the housebuilder might cash in on your lease and sell it to a speculative broker, who may want ten times what the initial housebuilder led you to believe would be the purchase price of your leasehold. That's what you have to be worried about.

This scheme/scandal has been called the PPI of the housing industry. I don't disagree. The PPI scandal was another mis-selling saga. This one is just the same but with a different name.

Thankfully, this dreadful system of selling something, yet keeping a bit back for yourself that you can sell on elsewhere, is being exposed and slowly wound down.

Michael Gove (Secretary of State for Levelling Up, Housing and Communities at the time of writing) had plans to outlaw the scheme in its entirety. Number 10 then scaled back on his proposals, deeming them 'too radical'! The Leasehold Reform Ground Rents Act of 2022 has now seen ground rents reduced to zero in most instances. Leaseholds have also been upgraded from 90 to 990 years.

So, this latest money-making rip-off fad has been outlawed, although only for properties sold after the act came into force. It doesn't provide for retrospective disbarment. The fact is, it should never have been allowed in the first place.

I really do hope that we have seen the back of this blatant exploitation of people who thought they were buying a home to live their lives in, but you better just check out the small print of your house purchase all the same. Be the king of your castle. Make sure it's yours, and that includes the land that it sits on.

THE COLLAPSE OF CARILLION

So, a building company went bust. Does it really matter? Well, yes. Certainly, in this instance.

Now, any company going under is a source of regret. When that company employs over 40,000 people globally, with over half of them based in the UK, then it's fairly significant.

According to a Sky News business report, Carillion went under for pretty much the same reason that all companies go bust: a debt mountain, with no one willing to lend them the salvation funds to get them out of the situation.

So, what went wrong?

Well, by the time they were placed in the hands of the Official Receiver on the 15th of January in 2018, the company had 450 government contracts on their books covering education, the prison service, major hospitals, housing, roads, the Ministry of Defence, you name it. They were builders, maintainers, and service providers across all of these construction sectors.

They were a big fish. Seen as a sure thing. A steady pair of hands.

According to the Assistant General Secretary of the Unite Union (of which more soon – and wait until you read about that!), the collapse and failure was nothing more than "rampant bandit capitalism". They also added that the UK's accounting and audit system was not fit for purpose because Carillion had been allowed to go careening off a cliff and was not brought to book until, well, such time as they were utterly defunct.

Their auditors, KPMG, one of the biggest auditing outfits, were fined £21 million and had to pay more than £5 million in legal costs – in relation to the "number, range and seriousness" of issues they had failed to address. Perhaps these so-called financial experts didn't understand what they were looking at? Or chose to look away.

Someone who did know what they were looking at, when they came to do their own due diligence, were those fellow big hitters in the construction world at Balfour Beatty.

A proposed merger between Balfour Beatty and Carillion in 2014 fell apart when the BB team cast their eyes over Carillion's order book and spotted some fundamental flaws. Namely, that the level of profit predicted on these contracts was rose-tinted, and the inherent risk involved in all their live projects simply wasn't being catered for.

As these were generally large projects, lasting several years in many cases, the margin for inflation, changes in the market, and events of a *force-majeure* nature were being undervalued or outright ignored. In essence, Carillion were going to make a very small profit margin on these schemes if everything went swimmingly and, adversely, a very large loss if things went awry. Which they often do in the world of construction.

Balfour Beatty decided to pass. Carillion's own supply chain started to get a little suspicious, too. Suppliers were placed on 120-day payment terms instead of the government guidelines (which no one takes a blind bit of notice of) of 14 or 30 days. When you're asking your suppliers to wait to be paid, you're basically saying to them, 'I ain't got no money'. You're simply buying time.

Meanwhile, in 2016, the chief executive of Carillion took home £1.5 million in salary and bonuses, with some of those bonuses paid as stock and options in the company.

And almost all of the senior management team, including the Chief Financial Officer, began selling their shares shortly before the company issued one of three profit warnings to shareholders in 2017. As they issued their first financial warning, Carillion wrote off a billion pounds of the profit they'd previously said they expected to make.

Never a good sign.

Still, the government continued to believe in them. Ignoring any adverse warnings, they awarded Carillion two HS2 contracts worth £1.3 billion just three weeks before the company went bust.

Labour MP Rachel Reeves has since said that the fault lay with the company's delusional directors, who drove the company off a cliff and then tried to blame everyone else. There were 70,000 people working within Carillion's supply chain, involving thousands of subcontracting companies, with many of them one-man or woman bands who needed the cash.

At a loss of £7 billion, affecting 70,000 people, that's £100K a person. Out of pocket.

Carillion's turnover at the time of collapse was £5.2 billion a year. They employed over 40,000 people. That means that every employee was generating about £125,000 a year. That's in sales.

From that, take the cost of the materials that you have to buy to carry out the job. Plus the company's own expected overhead and profit.

As a rough rule of thumb, allocate a third to the labour, the materials, and the overheads and profit. Overheads, in this instance, are the company offices, phones, internet, vans, marketing, etc, etc.

So, a third of £125K leaves around £41k per employee after deducting for materials and overheads.

Enough to pay staff wages, on average around £30K per annum.

Oh, but that includes the Chief Exec on £1.5 million a year.

Slightly diminishes the average.

The point here is that the margins are getting pretty thin. There's not a lot of room for manoeuvre.

But they're too big to go bust, right? Not a chance. As the old saying goes, 'Turnover is vanity. Profit is sanity.'

By this point in time, Carillion can't pay their bills and can't borrow the money to do so. It's a failure on so many fronts. They have not got the right staff with the right expertise to price the jobs correctly. They can't say no to a deal, even if it means slogging your guts out for five years and paying the client for the privilege, (i.e., not making a dime on the project).

I once heard the CEO of a very successful company say that they had come through the bad times of a recession by 'sticking to their knitting'. They knew what they were good at, they didn't jump on any bandwagon, and they didn't promise what they couldn't deliver. They simply said, 'This is us. This is what we do. If you want us, this is the price' because they knew they could make money and live to fight another day on those terms.

In other words, know your worth.

Carillion had good contracts – reliable contracts – Government work meaning taxpayers' money.

The only two things guaranteed in life are death and taxes. This was taxpayers' money. Therefore, it was guaranteed. But Carillion overstretched, took their eye off the ball, and as a result let everyone down.

Lessons to be learnt from all this?

No one is too big to fail.

No potential lender (or saviour) is attracted to a company running at a £1.5 billion annual deficit.

Read the warning signs.

Stick to your knitting, which for corporate enterprises and indeed for any business, means make sure that you're earning a profit on the things you know how to do well.

Otherwise, learn the art of saying no to contracts that you can't profitably deliver.

THE UNITE HOTEL FIASCO

Unite is the largest trade union in the United Kingdom, with about 1.4 million members. Subscription for full-time working members costs just £15 a month. It is even less for those who are only working part-time or for apprentices and other categories.

At an average of, say, £100 annually for those 1.4 million members, that's revenue of £140 million a year. That's not insignificant.

Back in the day, Unite decided to build a hotel and conference centre in the centre of Birmingham. It would be eight storeys high, with 150 hotel suites and a large function room with a thousand-seat capacity. The initial budget estimate, circa 2014, was reportedly £7 million.

By 2015, this had risen to £35 million, according to a leaked report viewed and reported on by *The Guardian* newspaper.

The contract for the work was awarded to a building company based 100 miles from the actual project. The company had no track record of constructing schemes of such magnitude, but they *were* close associates of the union's General Secretary.

In fact, that particular contractor moved into new offices themselves in 2015, premises which formerly belonged to Unite.

Anyway, after that first guesstimate of seven million, the estimate – increased to £35 million. A little while later, the estimate had ballooned to £57 million.

The scope of the project then changed. Just slightly.

Instead of the initial eight-storey building, they decided to go up to nine.

Let's do a bit of maths here to determine what an extra floor should cost.

A lot of the costs for any construction project harbour around the site set-up, management costs, etc. An extra floor will not come in at twice the price of the other floors; it will be about the same, or slightly less. So, if we are – by this stage of proceedings – at the £27 million estimate for the scheme (being kind and choosing the second highest of the four suggested sums), then let's add 10% to the budget for that extra floor. We are now near the £30 million mark.

Except the building ended up costing a total of £98 million!

Len McCluskey, the then-General Secretary (since ousted) explained with a straight face that it was because of the change of scope (no more than an additional ten percent of the budget by my reckoning) and the fact that Unite, as a responsible employer, employed best practices throughout.

He said, no doubt with clenched fist pumping the air, that Unite refused to compromise on the Health and Safety of the building workers employed on the scheme. They also wanted to ensure that every one of those heroic workers were paid a decent wage of at least the going national rates. Oh, and – apparently – construction costs in Birmingham are really high, he added.

Guess what. A bit of protective clothing (known in the industry as PPE) costs a pittance. It's a legal requirement, anyway. Anyone working in construction knows that you have to allow for it. Your estimator and your tendering contractors (if there were any) would include for it. A few hi-viz vests and hard hats won't send your costs up from £30 million to £98 million.

As for paying the workers a decent wage, you have to! No one will be turning up otherwise. Certainly not in the good times, which the years between 2015 and 2020 were… when the Unite hotel and conference centre was being built.

There is absolutely no traction whatsoever in these outlandish ideas. I think of them as excuses. The Unite members did likewise, and they voted Len McCluskey out in 2021.

He was replaced by Sharon Graham, who immediately ordered an independent inquiry into the whole fiasco to be led by a QC and an external law firm. Keir Starmer, the Labour Party leader, had called for the same thing earlier in the year.

No doubt the building has merit, but a recent valuation placed its worth at no more than £30 million. That's a huge waste of union members' fees. Many, including the new leadership, were aghast.

In April of 2022, police raided the headquarters of Unite in central London. Their investigation is concerned with bribery, fraud, and money laundering, all related to the awarding of the contract for that hotel.

Anyway, Len McCluskey appears quite happy with what he got for his members' £98 million. He wrote to *The Guardian* newspaper on the 20th of April 2021 in defence of the scheme and said that he believed that the development was a sensible investment.

I've no doubt that the building itself might be an asset, yet if its mortgageable value is only £30 million, I think that he considerably overspent. By about 300 percent.

McCluskey said, in that same tub-thumping epistle to that national newspaper, that his union's expenditure on the Unite hotel bore no resemblance to the systemic rinsing of the country that was going on in Westminster and Whitehall, and that it was ridiculous to claim equivalence.

So, there you have it. Case closed.

RAAC AND RUIN

At the end of August 2023, just one week before the start of the new academic year, 104 schools across the UK were forced to either fully or partially close their buildings due to concerns that they could no longer be considered safe.

Over the summer holidays, three instances of structural failure were found in a type of concrete panel used in the construction of many of our schools, hospitals, and other government buildings. It was a problem that had been aired much earlier – in fact, since the material's introduction in the 1950s – and now it was a problem that could no longer be ignored.

The suspect material at the heart of the issue is called RAAC. This stands for Reinforced Autoclaved Aerated Concrete. It is a lightweight material used largely in the flat roof construction of buildings, but which can also be used to form walls and intermediate floors. This means it can be right above your head, no matter which floor you are on.

And it's not very good.

It came with a lifespan of 30 years max and was used extensively from the 1950s to the 1990s. It was seen as a cheap alternative to traditional concrete. See, there's that word again. *Cheap.* Never ends well, does it?

So even buildings constructed in the early 1990s have now passed their use-by-date. Let's not beat around the bush. They are no longer safe.

Traditional concrete (which has a lifespan of around 100 years) is made up of cement (about 15%), water (about 15%), and aggregate (about 70%). Aggregate is essentially crushed stone and gravel. When the cement and water mix, they bind the aggregates into an impenetrable rock-like mass. Sounds impressive, right? And it is.

RAAC is made out of lime (nice as a slice), water (great in a desert) and an aeration agent (think Aero), which produces bubbles in the resulting material and fleshes it out.

So, yes, cheap.

Add a bit of steel in there, call it concrete, and away you go.

In 1961, not long after the material was first making headway in the marketplace, the Institute for Structural Engineers said that it was unfortunate that the term concrete had been retained for these aerated 'products'. Even exposure to short-term moisture reduced its strength by 13%, they said. Long-term exposure to polluted air reduced its strength by 40%.

And guess what? Those bubbles *will* let in moisture.

Ever heard the expression *leaking roof*? These prefabricated panels were largely used in the roof area of the buildings in which they were employed. So they are even more vulnerable to water ingress.

Once the water gets into those aerated bubbles (for which read 'gaps'), then it is going to come into contact with the steel mesh or the reinforced steel bars that are ingrained within to give the panels the strength that lime, water, and bubbles don't really provide.

Steel rusts. Then expands. And the pressure of that expansion causes the so-called concrete surrounding it to essentially burst, or combust, or otherwise shed integrity.

This fatal flaw has been known about for a long time.

In 1994, concerns about the use of RAAC were raised by the Building Research Establishment (BRE), a respected, formerly government-funded industry body.

Upon further investigation in 1996, excessive cracking and corrosion was found in many of the roof planks installed prior to 1980.

The concern was such that the UK effectively stopped using RAAC at the end of the last century, but we still have it in abundance in our current stock of buildings.

In 2002, the BRE stated, quite worryingly, that corrosion could occur in RAAC panelling without visual indication and that there was a risk of collapse *without warning* in panels over 20 years old. They went on to claim that many panels were inadequate and did not meet the regulations even at the time they were installed,

Since then, the Standing Committee on Structural Safety (SCOSS) has said that, though called concrete, RAAC is much weaker.

The Health and Safety Executive says that even panels that were installed in the 1990s were now beyond their lifespan and could collapse with little or no notice.

So, what is the government doing about it?

Well, too little, too late, in the opinion of many.

In 2018, a roof caved in at Singlewell School in Kent. The results were catastrophic. A complete implosion that left the roof in smithereens on the floor below. Luckily, the place was empty at the time as the incident happened on a weekend.

In 2021 and 2022, the Department for Education (DfE) sent out a questionnaire to all the schools in Britain, asking them to provide information on whether RAAC was present in their buildings.

These non-construction experts then began feeding back their answers. The advice at the time, issued by Nick Gibb, the Schools Minister, basically categorised RAAC as being in either a critical (i.e., anyone could see it was cracked, blown, and in every unsafe) or non-critical (meaning it looked alright to the naked eye).

Then, in the summer of 2023, there were three instances of RAAC failure, two of which occurred in schools in RAAC beams that had just been classed as non-critical.

The safest option, therefore, was to close every building in which RAAC was known to be present.

And that's where we are at.

The government's previous stance of fixing the problem at their leisure (endorsed by Rishi Sunak in his time as Chancellor of the Exchequer) is now no longer an option.

The funds will need to be found to replace these faulty (and life-threatening) products right now. Funds that will no doubt be purloined from some other needy department.

Schools, as well as those hospitals and the many other public buildings that employed the suspect RAAC material, will have to shut forthwith to either be demolished or to receive extensive structural enhancement.

That's the reality. Even if the dodgy beams weren't considered 'critical', the situation most certainly is.

14. RECENT INDUSTRY DEVELOPMENTS

GREEN TECHNOLOGY

We have one planet. Quite a nice one it is, too. In fact, I'd go as far as to say that it's the prettiest of them all. So, why don't we all do our bit to look after it? There are certainly things that we could all do to help. The world of construction – the same as every other area of industry – needs to do more. We need to look at our everyday activities, and all seek to improve the way we work.

We need to get inventive, creative, and think outside-the-box. Every aspect of a building's design needs careful consideration before a spade even enters the ground.

The first question we should ask ourselves is *why* are we building new? Could we not have refurbished something that already existed? Can we recycle any of the proposed or incumbent materials, possibly salvage and re-use them?

I'm convinced that in years to come (I'm talking decades, even centuries), we will be digging up our landfills and searching for the things we currently discard but which might still have a purpose in the future. So, why not start that recycling process now?

If we can make clothes and shoes out of discarded plastic, then what else might we be able to use it for? Single-use plastic is a scourge and a shame, and we are now seeing recycled plastics being used in flooring, cabling, pipes, and much more besides.

Of course, recycling need not just focus on obvious materials like plastics. In a home building project, for example, can we capture and store (and recycle) water that would otherwise pour down our drains? It rains, and the water disappears into our drainage systems, never to be seen again. Then, when the sun comes out, you reach for the hosepipe and give the garden turf and the border plants a good soaking. With fresh water from your taps. Well, why not capture that initial rainwater run-off in water butts or some other such container, then use that collection to water your plants?

That's an efficiency right there, and one which common sense tells us is a no-brainer.

The modern buzzword seems to be zero-carbon. Can your prospective development get there? If not, why not? Can you offset your damaging footprint any other way? Maybe plant a few trees in the landscaped gardens surrounding your brand-new building? That's an easy, sensible, and practical solution.

In fact, why not plant a few trees on your roof? Or at least a few plants. Green roofs and garden roofs are becoming increasingly prevalent in flat roof situations. They have one particular advantage in that they absorb far more solar heat than traditional flat roofs, which simply reflect the sun's rays back into the atmosphere. Green roofs also provide a haven for insects and birds, and bring the countryside into our ever-expanding urban landscape. A few years ago, this type of flat roof covering didn't even exist. Now, they grow (excuse the pun) more popular by the year and I, for one, am in favour.

Our homes and premises are all being asked, and regulated, to become more energy efficient. In the early 1980s, the UK went double-glazing mad. Homeowners felt compelled to have it installed. It saved on the heating bills and also reduced any noise travelling into and out of your abode.

Solar panelling, by comparison, hasn't quite taken off in the same way domestically. For one thing, they look unsightly. While the market continues to manoeuvre to more aesthetically pleasing solutions – such as integrated, tile-effect panels that literally won't stick out as much – they are still a hard sell. They're not yet universally popular. They've not yet had their day in the sun, as it were, but it will probably come.

Still, solar panels seem to have found a burgeoning opportunity on solar farms, where row upon row of the things can be lined up on vacant fields where they generate huge amounts of electricity that can then be sold to the national grid.

As the recent conflict in Ukraine has shown, our energy and food supplies can be compromised by global events. The more locally you can generate the things that are essential to our daily lives, the better it is for all concerned.

We now clamour to insulate our homes. The government even offers grants to homeowners to help them along the path to thermal efficiency. Less heat escaping through your walls and your roof means less of the hot stuff entering the atmosphere and ultimately warming the planet.

There are so many ways in which the construction industry can improve the way it works as we seek to embrace the challenge of becoming greener and protecting the planet. The world of construction, like everything else, simply cannot ignore the harsh reality.

It's time for us all to put on our hard hats, safety boots, and our hi-viz jackets and get to work addressing these important green issues – big, planetary issues – before it's too late for us all.

BUILDING INFORMATION MODELLING (BIM)

Building Information Modelling, or BIM for short, has been heralded as the future of the building industry. One day, it is hoped all construction schemes will be run on its platform. It is the bright new dawn of our bricks-and-mortar industry. The Neanderthals amongst us are about to get left behind.

Or so the story goes.

Personally, I don't believe it. Though I welcome anything that improves our lot, some things won't ever really change.

The idea of BIM is to create a geometric 3D model for every new envisaged building, that incorporates each element of its design. This can then be shared with everyone involved in the project. All the stakeholders – from the architects, the product manufacturers, the mechanical and electrical engineers, and the contractors and subcontractors – can contribute to it.

The result is a perfect picture of how the planned building is meant to look and operate, so that even the eventual occupiers can understand the nuts and bolts of the structure in order to problem-solve and maintain the building in the coming years.

Now, personally, I'm a bit of a technophobe. I'm writing this book on an old typewriter for one thing (only joking!), but I am definitely old-school when it comes to the use of and the understanding of technology. For example, I value the old-fashioned tradesman's apprenticeships that last for several years rather than the six-month courses which claim to turn you out as a bricklayer or some other such practical skill in double-quick time. I wouldn't let anyone build *my* house with only a six-month certificate behind them; I know that much.

I recently walked around a new construction project with an architect who held a tablet that had all sorts of 3D images appearing on screen at the touch of his hand. It was quite the space-age thing. This was for a relatively modest scheme of turning a large derelict house into an eight-bed HMO. I was really impressed. But I didn't use the technology myself, nor did the planner, and nor did the builder. Universal, it is not.

The company I currently work for don't appear to have any plans to get involved in BIM. I've not had any experience with it, so I'm probably not the best person to ask. But if you *were* to ask me if I considered this tool to be the future of our industry, I'd have to say no. Not for a while, at least. Most people I know who work in the industry don't use it. It's just another box-ticking exercise.

Although tendering contractors are now told that it's a requirement for many large public projects, it's a problem that is easily overcome. Buy the software and send the client a few images. Then, go back to working the way that you always did.

I've seen comments online that BIM will render certain occupations in the industry redundant, such as quantity surveyors. The amazing computer, having drawn the building from every angle possible, can then extract the quantities of materials contained in the architect's drawings. This will, therefore, negate the expertise and manhours of QSs, estimators, and presumably many other construction experts.

Us surveyors will all soon be out of a job, apparently.

Well, when a client doesn't pay your interim payment application – the money from which you need to pay your subbies and your own overheads – is BIM going to track the bloke down on the golf course and give him an ear-bashing and demand that he hand over your dough? Don't think so.

So, great idea. Great visuals. Probably going to be much more prevalent in the future. For now, the *real* world of construction still wins it for me. At this moment in time, BIM is in the bin.

OFFSITE PRODUCTION

Ever since the Egan Report (which gave an overview of the British construction industry in 1998), there has been a clamour to utilise more offsite production within the building game. The underlying message being, "Get into the 21st century. Offsite manufacturing is the way to go." And by offsite manufacturing, I am talking about pre-fabricated units such as wall panels that come with pre-fitted electrical sockets, roof panels that come with pre-fitted tiles, and even modular homes that contain pretty much everything, pre-made in a factory, shipped to site, then connected to the amenities.

I'm not against it. Some of these products are pretty impressive, but don't you just love all of these Government initiatives? Words on a paper that are somehow meant to resonate, educate, and make everybody see the light.

For the majority of people plying their trade in the industry, the new music emanating from these ivory towers simply falls on deaf ears. Still, I recognise that there will be a gradual shift in this direction, to offload some construction activities into the realm of manufacturing. I understand the reasons why, yet there won't be an overnight change in the way that things have previously been done.

Opportunities will be explored, and if they can prove their worth, they will be adopted. Otherwise, they'll simply fall by the wayside as a result of market forces and practical onsite realities.

Let's face it. The construction industry is a huge consumer of manufactured goods. Bricks are made in factories. As are roof tiles, glazing, doors, steelwork, and curtain walling. Add to this pre-cast concrete or timber stairs, bags of plaster, and paint. All of them made offsite, then delivered to site, and installed onsite.

The system works pretty well as it is. And, if anyone ever does come up with a cost-effective labour-saving product that clients are happy to accept, then you can bet your better dollar that it will soon become the norm.

There are pros and cons to offsite production. In the plus column, offsite manufacture should mean uniform quality. But does it? Are factory workers any more skilled than construction workers? Ever bought a faulty product? Things can roll off a production line in less-than-perfect shape. Even a new car can have faults. Then it's back to the production line.

Another plus is that, once delivered to site, pre-made products take less time to install. You're simply fitting the thing. This presumably means a safer working environment because the physical work then gets done much quicker, and workers spend less time on dangerous construction sites.

It's true that construction is a more dangerous place to ply your trade than manufacturing, but not by all that much. There were 30 fatal injuries in construction over 12 months between 2021 and 2022. There were 22 in manufacturing during the same period. And try telling that story to the poor sod (on a building project I was involved in) who got hit on the head by a modular unit as it was being craned into position!

So, while manufacturing is safer, that doesn't mean it is without risk. There is still an interface between the materials produced offsite and those onsite who install them.

Another supposed positive with pre-made is that there is less waste. That makes sense. If you've paid for something to be delivered to site, then presumably you want it in its entirety and you don't want to discard any of it. I'll certainly go along with that. It can also work out cheaper to construct. Make it offsite in a factory environment, bring it along, bed it down, and plug it in. Bob's your uncle. But only if you buy a lot of it.

The cost of creating the initial template, mould, or whatever – that you can then run numerous examples of – rarely comes cheap. It takes time and costs money. It is simply not worth the effort unless you're going to produce a number of these items.

The main thing about ordering pre-fabricated items is that you have to get your design absolutely right at the outset. It takes planning and forethought, coordination and understanding, as well as liaison and clear and unambiguous communication with the manufacturers. What is their lead-in time? When can they get the goods you want (that you've clearly detailed and described), in the quantity you need, produced and delivered?

So, while there may be some savings on time spent onsite, you'll have probably expended the majority of that in the pre-planning and pre-ordering phase. You'll likely end up in about the same place in terms of your overall project programme.

And there are also logistical issues involved with offsite production too. Big lorries turning up at regular intervals with big loads, that then need to be lifted off the back of those articulated trucks and hauled into place, takes a bit of lateral thinking. Can your building site take possession of such deliveries? Do you have the swing radius to lift those products into place? If not, it's back to the traditional methods you go.

Other possible downsides to offsite production include the lack of choice in the current market. As such, only 2% of current industry turnover is spent in the offsite arena.

And what about late design changes? If the client or architect has a change of heart, who is going to communicate that to the factory owner who has already rolled 50 of your first-choice items off the production line?

Prefab is a name we associate with the 1960s and 1970s, but does it hark back to a golden age of construction? I think not.

When I think of prefab homes, I think of buildings thrown up in times and areas of dire need when almost anything would do, like at the end of the Second World War. In turn, you hear tales of how mortgage lenders shy away (and simply refuse) to back non-standard constructions (think social housing, for example).

Modular homes are improving, it must be said, pioneered by firms based in places like Scandinavia and Germany. Indeed, Huf Haus properties are expensive and beautifully made.

On one such modular development, at an army camp, I saw prefabs being lifted into place. Internal walls, electrics, kitchens and bathrooms were all in situ and – floor after floor, and side by side – these units were stacked into place.

What did the contractor then do?

They built the brick exterior, put roof trusses on it, and put roof tiles on those trusses! So, even when you have a modular scheme, the traditional construction trades still abound.

I once worked on a housing project where we purchased pre-fabricated timbers that would form the inner core of the building – essentially the beams, joists, girders, and trusses. All made offsite; all delivered pre-made and ready to be lifted into position.

They were great.

What wasn't so great was the concrete floor slab that they were meant to sit on. The site manager had built the slab in the wrong place! He also blamed the architect. The thing about any enterprise that involves so many professions, so many trades, and so many interested parties, is that there's always someone else to blame!

In this case, they got the prefab bit right. It was everything that went before, the supporting construction, that was horribly wrong.

My point here surrounds scale. If you're building the channel tunnel, you can design a concrete section, say 20 metres long, and replicate that a thousand times (or more like 25,000 times for a 30-mile-long tunnel). That makes sense.

As does manufacturing bricks, making paint and plaster and roof tiles in factories, and cold-rolling steel, and all of the other things that already happen where construction and manufacturing connect.

Where prefabs can often come unstuck is when you're only producing items in small numbers. It's more cottage industry than production line.

The future of the industry? I think not.

Gradually, slowly, more products will no doubt come online with improvements in quality, cost and minimum numbers. But for any brickies, plasterers, painters, joiners, sparks, and other construction industry trades reading this book, I don't think we'll be seeing the back of you anytime soon.

HEALTH AND SAFETY

I think it's fair to say that one of the most visible advances in construction over the last 50 years has been in the industry's approach to Health and Safety. While still just about the most dangerous arena in which to work, it is certainly a lot safer than it was a half-century ago.

Fatal accidents in 2020 accounted for 30 lost souls, which is a tenth of the number who lost their lives in the building trade annually 60 years earlier, so we have come a long way, though it is still the most dangerous industry, followed by Agriculture (including Forestry and Fishing) and then Manufacturing.

Legislation has brought about a lot of this positive change. From the Health and Safety at Work Act (also known as HSWA, the HSW Act, or HASAWA) of 1974, through to the current Construction Design and Management (CDM) rules that govern our industry. No doubt being part of the European Union (while it lasted) also helped the UK to up its game.

And why not? We can't have our hardworking employees and colleagues taking their lives in their hands every time they turn up to a job. As one of our black hat leaders (different coloured hard hats represent different roles; black is for site supervisors) used to drill into us on one prestigious project I was involved in, 'None of us gets paid enough to risk our health and safety.'

In fact, it is almost driven into every construction operative nowadays that Health and Safety is the most important aspect of every project. It is the one area where no corners can be cut.

You can skimp on costs, on design, on the quality of materials and more (as long as you don't substitute safe for dangerous!), but Health and Safety is the great untouchable. Which is exactly as it should be.

Onsite, each worker usually wears three or four points of PPE. That's Personal Protective Equipment. Your most obvious are your hard hat, steel toe-capped boots, and a high-visibility vest or jacket. Then comes the protective goggles. Ear plugs, too, in noisy work environments. Face masks covering the nose and mouth are required in dusty work situations. Sometimes, we even have hi-viz trousers alongside our hi-viz vests to fully round off the Dipsy (*Teletubbies*) appearance.

Before a subcontractor is even allowed onsite, they must provide RAMS for each work activity that they are scheduled to do, and RAMS stands for Risk Assessment and Method Statement. In this, you basically have to say what you are going to be doing, and how you plan to carry out all of the tasks involved in a safe manner.

You are principally providing documentary evidence that you've assessed the risks involved, and you are providing a statement of the methods you are going to use to carry out those tasks in a safe way. It shows that you've thought about the work ahead and the hazards you might meet along the way. And, having thought about them, that you have mitigated or nullified any potential ramifications. Not a bad thing, to think about what you're going to be doing and the dangers that might then exist.

So entrenched is this constantly improving way of working, for everyone involved in construction, that it is now common to see examples on social media of 'Idiots at Work' and other such handles.

I can remember a time 30 years ago, when I was little more than a humble labourer, how we would re-roof houses off a ladder! If we had any form of scaffolding on a job, it was an absolute luxury.

I would bump tiles up a ladder on my shoulder, and the roofer would take them off me and walk up the wooden battens and stockpile the tiles on the roof before laying them. If he hadn't been as comfortable as a cat up there, he could easily have lost his balance, and there would have been no fall-arrest or crash deck to stop his descent to the ground. In all likelihood, with fatal consequences; falling from height is the biggest cause of current construction-related deaths or serious injury.

Obviously, these improvements in our working conditions have come at a cost. Not a human cost but an economic one.

Scaffolding is a big one. No one works off ladders any more. We now have working platforms and edge protection. If someone tried to win a construction job by cutting corners (maybe foregoing necessary scaffold costs, or proposing to fall short on any other Health and Safety matter), then their tender would simply be deemed non-compliant and into the bin it would go.

In turn, company directors know that they can ultimately be held responsible for accidents or deaths in the workplace that are caused by a lack of statutory safeguards. These can result in fines and even jail time for negligent parties, so sense has prevailed across the length and breadth of the industry.

In 2023, we have seen some of these lofty people jailed, given suspended prison terms, and extensively fined for H&S failures that have seen staff injured and, in some cases, killed.

According to a report in *Construction News* in May 2023, two directors of a building firm were jailed, and their company fined £1.6 million after five workers were fatally injured when a 45-tonne wall fell on them at a site in Birmingham.

Many building firms now employ their own Health and Safety officer to monitor their working practices. It gives them a chance to keep their own house in order before the dreaded site visit from the HSE, which could result in your building site being shut down, although one Health and Safety officer once said to me, 'The boss sees my salary as dead money.'

In other words, he wasn't actually contributing to the boss's bottom line. He wasn't laying bricks (even though this fella was a former time-served brickie) or putting paint or plaster on the walls. He was just a required overhead on the company's payroll. That was how the boss saw it.

I'd beg to differ.

I think these guys do a very important job. When you're working at the coal face, if you're working on a price-job, you can get a little snow blind and go full tilt at the task in hand. You lose sight of what's going on around you. So, to have someone looking out for you, looking over your shoulder, checking the surrounding environment for hazards, well, that can only be a good thing in my book.

So, while our construction projects are now inflated with the added costs of scaffold, welfare facilities, PPE, and the salary of an in-house safety officer, the number of fatalities and those suffering serious accidents onsite have decreased considerably decade on decade. By 90 percent.

We no longer traverse the muddy and dangerous building site in our trainers, in danger of standing on a protruding nail, or risk having our toes broken (thanks to our steel-capped boots) or our heads cracked open by a bit of falling masonry or a misplaced roof tile (because we've got hard hats). We're safer.

We get to go home each night in one piece, much as we left in the morning. This is one area in which the construction industry, in recent years, has most certainly improved. In fact, it could even be our greatest development of the last 50 years.

15. THE FUTURE OF THE CONSTRUCTION INDUSTRY

What does the future look like for the construction industry? Does it even have one?

Well, the good news is that the answer seems to be a resounding yes.

Construction activity is expected to grow by 47% in the decade from 2020 to 2030. This is in terms of both the personnel involved and the output it delivers.

In a nutshell, we simply need more houses and more infrastructure to meet the demands of a growing population.

The number of people required to help build and then maintain these assets is also set to grow. Recruits need to be quickly found, trained, then set loose on that growing list of pending projects.

We need investment in the training, skills, and crafts that are required to ensure we build products of suitable and sustainable quality.

Our roads and railways constantly need refreshing.

How about that levelling-up agenda between the north and the south of the country?

Whenever the national budget gets squeezed, the south seems to keep their bit of the development – as with the HS2 network – while the north gets to make do and mend with the odd improvement here and there. The sounds of their disapproval never seem to reach those far-away ears in Westminster.

As we move towards that fast-approaching future, none of us can be sure how it will really look or work.

I think it's fair to say, though, that people will always need accommodation and factories, shops and supermarkets, warehouses and retail places. All of these will need decent connectivity via the roads, the railways, the ports and airports, as well as the bridges and tunnels that transport us all from A to B.

We'll still need our water supply. We'll still need to dispose of our waste.

There is huge room for improvement in the way we 'do what we do' in our daily lives, but the things we actually do are unlikely to change all that much in the future.

So, new ways of doing the same old things. That's what we are looking at.

And let's not forget the basics.

Build dams! Capture water. We can't live more than three days without it, so why risk going without?

Prisons? We might need more of those in the future. The same with hospitals and care homes for an ever-aging population.

After that, I'm not prepared to say. After all, this is a book. It's not a crystal ball!

There will, of course, be economic challenges along the way. The cycle of boom and bust seems set to continue. No one appears to have found the solution to it yet in the annals of capitalist history.

In the boom times, the working population can exploit the demand for their services and secure higher wages. In the bust, we'll all eat each other in the quest to be the last man standing.

But even if (and when) we do return to the dark times of recession (which could be months or weeks away in our current economic uncertainty), there will always be something better down the line. The clouds will part and the sun will eventually come out and shine again.

I believe that construction, in general, is ready for the future. We either have or we can certainly learn the skills necessary to meet the challenges that lie ahead.

And the biggest of those challenges may well be that of climate change. When parts of the world get too hot to handle, or flood regularly, the populations in those affected areas will have to migrate. Wherever they end up, there will then be an increased need for housing and supporting infrastructure.

That's the future of the construction industry secured! Putting the world to rights.

CONCLUSION

So, there you have it. The UK Construction Industry, as told to you by The Secret Surveyor.

This is the sort of stuff that the RICS won't be telling you any time soon. This is the warts and all building game. Who it employs, what it does, its history, and its future.

We've looked at different types of project, and the different sectors, large and small. We've also looked inside and outside of your home.

Hopefully, you now know how to make and save money with your property investment, or when you come to engage with a building contractor in daily life.

Having read this book, you should sound like you know what you're talking about going forward. That's always handy when it comes to not getting ripped off.

We've met the industry's professionals and tradespeople, and discovered what they do, what it is that they do well, and what they sometimes do badly.

We've looked at project costs. How to price your own work, and how to ensure you're getting value for money.

We've also discussed some recent construction disasters. The resultant misery can affect us all.

And, finally, we've discussed some of the challenges that the construction industry is currently facing and will face in the decades to come. We've been talking about our built environment.

I hope that you see the buildings in your life in a slightly different way now that you've learned how they have been designed and constructed, and how everything is connected.

It is truly important to each and every one of us. It's not just bricks and mortar; it's the fabric that surrounds us.

That's the UK Construction Industry.

I've been your guide.

I am The Secret Surveyor.

Other books from the publisher

The Secret Magistrate

Every criminal case starts in a magistrates' court, and most end there. Last year, the 14,000 magistrates of England & Wales dealt with almost 1.4 million cases.

But, what exactly does a magistrate do, who are they, and how are they recruited and trained? Are they out-of-touch and unrepresentative, or still fit for purpose with a role to play in today's increasingly sophisticated and complex judicial system?

The Secret Magistrate takes the reader on an eye-opening, behind-the-scenes tour of a year in the life of an inner-city magistrate. Chapters cover a variety of cases including the disqualified driver who drove away from court, the Sunbed Pervert, and Fifi the Attack Chihuahua.

All In Your Head: What Happens When Your Doctor Doesn't Believe You?

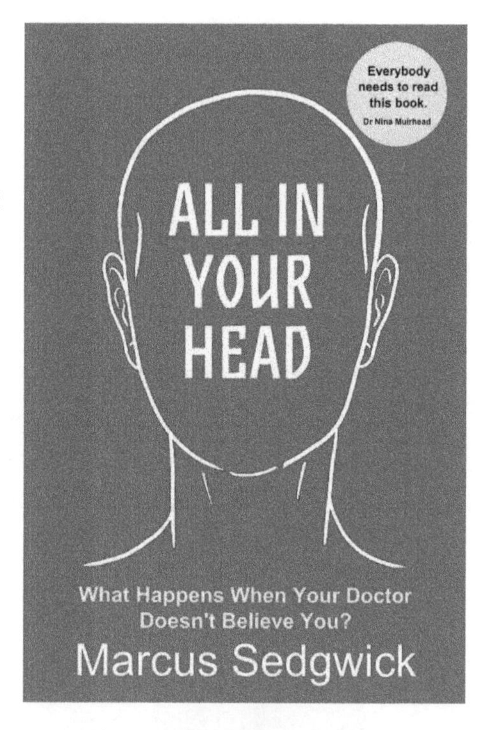

All In Your Head is about what happens when your doctor doesn't believe that you're ill. When they think you are imagining a serious ailment, or worse, faking it.

It's the story of the stigma that goes with invisible illness, and of the strange places that chronic illness takes you. It's the tale of bizarre treatments, and above all, the damage that's created through other people's doubts and indifference.

Yet, there is an epidemic of undiagnosed, hard-to-explain, and misunderstood illnesses in today's world, with new illnesses such as long-COVID steadily emerging. It is often up to individuals to drive their *own* search for recognition and a diagnosis, a task that can prove challenging due to establishment scepticism and disinterest.

With honesty, and at times, dark humour, *All In Your Head* – from multiple award-winning author Marcus Sedgwick – explores how four simple words can make you question your sense of reality.

(Out)Law: From Teenage Mum to Legal Trailblazer

(Out)Law is a powerful and inspiring journey of survival and resourcefulness, exploring the remarkable life of Alice Stephenson, who defied adversity to become a beacon of change in the legal world. At just 18, Alice faced the challenges of teenage motherhood and homelessness, yet refused to be defined by her circumstances. Determined and focused, she embarked on a path of education, navigating the complexities of university while juggling her responsibilities as a young mother.

Having qualified as a solicitor, Alice had proven that barriers could be broken down thanks to grit and passion, but her story doesn't stop there. In 2017, she set out to disrupt the legal industry by founding her own law firm – Stephenson Law.

(Out)Law is a testament to one person's defiance of societal expectations when coupled with drive and ambition. This book explores the struggles and challenges that women and those whose faces don't fit the mould encounter, as Alice stands against the norms of a male-dominated, elitist legal profession.

Are We Still Rolling? Studios, Drugs and Rock 'n' Roll – One Man's Journey Recording Classic Albums

Phill Brown is the sound engineer and producer who – over an illustrious 50-year career – has worked with many of the biggest names in rock and roll.

In this reissued and updated version of his 2011 memoir, Phill describes the ups and downs of a professional recording studio, working on sessions for The Rolling Stones, Jimi Hendrix, and Joe Cocker at the famed Olympic Sound Studios in London. As a young sound engineer, Phill learned the ropes from experienced engineers and producers such as Glyn Johns and Eddie Kramer, and soon worked his way up the ladder, engineering sessions and producing albums. His remarkable roll call includes Steve Winwood, David Bowie, Led Zeppelin, The Rolling Stones, Jeff Beck, Pink Floyd, Bob Marley, Talk Talk, Roxy Music, Go West, Dido, and many other legendary rockers.

With a foreword by Robert Palmer, *Are We Still Rolling?* is more than a recollection of treasured music-making over 50 years. It is one man's journey through a life where drug abuse, chaos, rampant egos, greed, lies and the increasingly invasive record business all took their toll. It's also a cautionary tale, where long workdays and what once seemed like harmless indulgences had sinister consequences.

Milton Keynes UK
Ingram Content Group UK Ltd.
UKHW020018130424
441043UK00006B/147